Advances in Underwater Technology, Ocean Science and Offshore Engineering

Volume 8

Exclusive Economic Zones

Resources, Opportunities and the Legal Regime

ADVANCES IN UNDERWATER TECHNOLOGY, OCEAN SCIENCE AND OFFSHORE ENGINEERING

Vol. 1 Developments in Diving Technology
Vol. 2 Design and Installation of Subsea Systems
Vol. 3 Offshore Site Investigation
Vol. 4 Evaluation, Comparison and Calibration of Oceanographic Instruments
Vol. 5 Submersible Technology
Vol. 6 Oceanology
Vol. 7 Subsea Control and Data Acquisition
Vol. 8 Exclusive Economic Zones

The conference was planned by Don Lennard, in co-operation with John King, Roger Lowndes and Geoff Larminie.

Advances in Underwater Technology, Ocean Science and Offshore Engineering

Volume 8

Exclusive Economic Zones

Resources, Opportunities and the Legal Regime

Proceedings of an international conference (Exclusive Economic Zones—Resources, Opportunities and the Legal Regime) organized by the Society for Underwater Technology, sponsored by the Society for Underwater Technology, the Royal Geographical Society and the Royal Institute of Chartered Surveyors and held in London, UK, 20 May, 1986.

Graham & Trotman Ltd.

A Member of the Kluwer Academic Publishers Group

First published in 1986 by
Graham & Trotman Limited
Sterling House
66 Wilton Road
London SW1V 1DE

Graham & Trotman Inc.
101 Philip Drive
Assinippi Park
Norwell, MA 02061
USA

British Library Cataloguing in Publication Data

Exclusive Economic Zones — Resources, Opportunities and the Legal Regime
(Conference: 1986: London) Exclusive economic zones — resources, opportunities and the legal regime: proceedings of an international conference (Exclusive Economic Zones — Resources Opportunities and the Legal Regime) — Advances in underwater technology and offshore engineering; v.8)
1. Marine resources
I. Title II. Society for Underwater Technology
III. Royal Geographical Society IV. Royal Institution of Chartered Surveyors V. Series
333.91'64 GC1015.2

ISBN 0 86010 774 4 (hardback)
ISBN 0 86010 441 9 (Series)

Printed in Great Britain by
Antony Rowe Ltd., Chippenham

Contents

The following paper presented at the conference was not available for publication:

Remote sensing from space — surveying the resources of the EEZ,
R.H. Thomas

Foreword

The topics covered in these proceedings are of the highest importance both nationally and internationally. They concern recent national and international developments with regard to marine resources and management of the oceans. They cover a broad range of subjects, but in one way or another they are related to the subject interests of our three sponsoring societies: the Society for Underwater Technology, the Royal Institution of Chartered Surveyors, and the Royal Geographical Society. These conference proceedings will encompass:

- underwater and marine technology and science;
- surveying, terrestrial and marine, and the uses made of survey information;
- the broader geographical, legal and geopolitical aspects of the expanding knowledge of our planet and its role in human affairs.

The ideas and issues surrounding Exclusive Economic Zones are important in many ways. They are important professionally: they lead us into the frontiers of technology, of marine science, biology, geophysics and environmental knowledge; of international and domestic law and economics, and the theory and practice of policy formulation and management of resources. They are also important politically, from both a domestic and an international aspect. Contributors to these proceedings represent various countries and jurisdictions that have taken quite different approaches to this problem which, like few others in recent decades, focuses on the politics of management of our planet, its resources and its environment. The issues are also important economically: from the point of view of the exploitation or preservation of present resources; of assessing profitability and how to maintain it; of the need for substantial public and private investments; of the costs of environmental degradation or preservation; and of the development and application of commercial or subsidized technology into new areas of activity.

But over and above these factors, the ideas surrounding Exclusive Economic Zones are important conceptually. The ideas are more difficult and complex than they first appear, and are controversial in a philosophical and cultural sense as well as in the scientific, political, and economic. To some, the evolving ideas are forward-looking, and the developments are inevitable in an increasingly technological and interdependent yet competitive world. To others, the recent developments are retrogressive, a throwback to more primitive times, perhaps the last kick of an anachronistic nationalism that is self-defeating and which cannot be sustained, politically or economically.

As men and women sincerely interested in this subject, we must not close our ears to any side of the various issues, nor look upon Exclusive Economic Zones as a subject in itself. The concept of Exclusive Economic Zones, and the rapid trend towards their declaration around the world, make a dramatic, significant example of the attempts of an increasingly populous and technological, economically uncertain and politically fragmented society to manage its institutions and its resources on a finite planet.

E.F. Roots
Conference Chairman

Exclusive Economic Zones: keynote address

Lord Sherfield

The subject of Exclusive Economic Zones is important, and it is topical. In a nutshell, about 36% of the total area of the ocean falls within the EEZ. Of the 139 maritime nations in the world, 103 have declared their EEZs unilaterally. The UK has not declared its EEZ, though it does claim an exclusive fishery zone around the UK and Rockall amounting to 274,800 square nautical miles.

It is estimated that EEZs worldwide account for about 90% of the present commercially exploitable fish stocks, 87% of known submarine oil and gas desposits, and about 10% of known polymetallic sulphide deposits, as well as large deposits of sand, gravel and other useful elements.

In my capacity as Chairman of the House of Lords Select Committee on Science and Technology, I took part in an enquiry into the state of Marine Science and Technology in the United Kingdom. Reports of these Select Committees are made, not to the public at large, but to the House of Lords, by whom they are appointed. There is then a debate in which a Minister makes a preliminary reply on behalf of the Government. Then, usually after an interval of two or three months, the Government replies in writing to each of the Committee's recommendations individually. We are more than half-way through that process. The report was published on 18 December 1985, the debate was held on 7 April 1986, and we expect a definitive reply to our recommendations before the end of July.

The report laid special emphasis on the national and international importance of EEZs, and in addition, the following points were made.

The Committee noted that at present there was no coherent determination of future research priorities, and considered it important that a research strategy should be developed and should be kept under constant review. An essential element in this strategy should be the provision of a scientific basis for the investigation, wise use, and management of an EEZ. We noted that the UK had not yet declared a full EEZ, let alone developed a policy for its comprehensive management and exploitation. This would require an improvement of the fundamental knowledge of the physical, chemical and biological processes of the seas and development of a scientific basis for the management of the marine environment. The Committee found that little integrated research had hitherto been devoted to these objectives, and suggested that a major aim should be to develop techniques and technologies for use in the exploration and exploitation of foreign EEZs, since there are potential overseas markets for such work, notably in former Commmonwealth countries and among the 66 African, Caribbean and Pacific nations who are signatories to the Lomé Convention. The report identified three main areas where research was needed and where it would promise important rewards : environmental protection, resource evaluation and underwater engineering. Finally, it was recommended that a full resource evaluation study of the UK EEZ should be mounted to demonstrate to overseas customers the excellence of British technology.

In the debate, Lord Cranbrook and I laid special emphasis on these and related points about the EEZ, but I am sorry to report that the Government reply was very negative, apparently on the grounds that we already had an extensive fishing zone which did not have to be declared. This really missed the point. The Minister was, of course, expressing a preliminary view. Hope springs eternal, and perhaps the Government in their written reply will have reflected further on our views on this matter, and will come up with something positive and constructive. Prudent considerations have not prevented our competitors from declaring their EEZs and actively surveying them, in the case of the USA with the aid of the British-designed and made side-scanning vehicle "GLORIA".

More widely, the Committee considered that projects which were concerned with the deep and distant seas would complement research directed to more immediate practical and commercial ends in the EEZ. However, they recognizedthat the expense of such research precluded the UK from continuing any extensive programme of research on a unilateral basis. International collaboration was essential. The Committee therefore welcomed the long-delayed decision to join the second stage of the ocean drilling programme.

While indisputably valuable in itself, involvement in such international research programmes also has practical repercussions on EEZ research. It raises the UK's profile on the international stage and helps establish what the Foreign Office termed "the reputation of the UK as a reliable partner in international science". When other countries wish to obtain assistance in evaluating the resources of their own EEZ they will turn to countries with which they already have some familiarity. A UK presence in distant seas, and some experience of the scientific and logistic problems to be encountered there, would be advantageous. And, of course, participation in international programmes broadens the range of scientific and diplomatic contact between nations. This is a definite advantage in view of the increasing difficulty of obtaining clearances for research vessel cruises in other countries' EEZs.

Finally, the Committee were firmly of the opinion that research in such a dynamic medium as the ocean could not be artificially confined by a mathematically-derived boundary line. As the Institute for Marine Environmental Research put it, "Oceanography as a science is certainly global in its reach and in its ambitions; to constrain this reach is to damage the science".

It is with this wider canvas in mind that the issues raised by the advent of Exclusive Economic Zones should be considered.

Exclusive Economic Zones - A Brief Sketch of Historical Development and Current Issues

E.F. Roots, Department of the Environment, Ottawa Canada

Background

From the earliest recorded development of ideas of organized territory, and at least from the times of the Phoenecians (ca 1300 B.C.), it appears that parts of the seas "belonged" as did the land, to whichever military or political powers could control them. Whenever there was power to exert control, the principle of mare nostrum was in effect, and was applied, when enforceable, by successive powers in the Mediterranean, in the coastal waters of Arabia and China, and in the Baltic and the Atlantic by the Norsemen (see for example Blom 1984). A practical if primitive version of Exclusive Economic Zones was taken as normal by governors and governed.

Although Roman law had taught that the sea was common and for the use of man (maris communen usum omnibus hominibus), responsibility for protection (from piracy) and jurisdiction (for taxation and right of salvage) was assigned to the "master of the sea", i.e. the head of state possessing or governing the adjoining land area. The principle of mare nostrum did not extend very far, but was restricted to seas where power could be exercised and which were considered important, for resources, trade or military purposes. The areas beyond mare nostrum were mare vastum, the uncontrolled or outer seas, beyond jurisdiction but not in any way considered high seas or free seas as we understand them today. The limits of national seas were taken as the half-way points between coasts or territorial jurisdictions in early Icelandic-Norwegian law, and in 1023 the median line principle was evoked by King Canute of England in dividing the sea between England and Gaul. By that time, presumably, marine surveyors had a job to do to define exclusive maritime zones.

The development of competing sea powers in Europe - Norway, Denmark, England, Holland, Portugal, and Spain - led to a breakdown of the limited mare nostrum concept and the evolution, among rival powers, of the principle of mare clausem, or sea under exclusive sovereignty of a responsible state. Many declarations expressing the mare clausem principle were issued from the thirteenth century on, and this principle, and the challenges to it, have lain behind much of the subsequent course of world history. From a Papal decree in the twelfth century, Venice had "dominion" over the Adriatic, and this formed an effective barrier to the westward expansion of the Turkish Empire. Venice claimed an Exclusive Economic and Political Zone. The Norwegian Royal Ordinance of 1294 made it illegal for foreign ships to sail north of Bergen, and similar decrees by other members of the Hanseatic League effectively carved the northeastern Atlantic Ocean into a number of exclusive maritime zones. The decree of Pope Alexander VI Borgia of 1493 followed by the Treaty of Tordesillas in 1494 was more ambitious. The Treaty capitalized very quickly on Columbus' discoveries by establishing a line, running from the North Pole to the South Pole and passing 370 leagues west of the Cape Verde Islands, from which not only were all lands to the west and not previously in the possession of a Christian power to belong to Spain and all similar lands to the east to belong to Portugal, but no vessels or representatives of other states were allowed to voyage to the new lands, or to fish, without royal license and authority. These and similar such declarations, enforced by powerful trading leagues, naval policing, or sanctioned by the "international" church, established a widespread regime of acceptance of the principle of national control over marine resources and activities in defined areas of the oceans (Fulton, 1911).

One of the most dramatic examples of the principle mare clausem in a more or less modern context of economic control was the ukase of Tsar Alexander I, issued in 1821, which closed the North Pacific Ocean. Under this ukase, foreign ships were prohibited from approaching within 100 nautical miles of the coast in an arc extending from Vancouver Island in North America, around the Bering Sea to Japan. Any ship seized in this huge zone was to be taken to Petropavlovsk in Kamchatka, where its cargo would be confiscated and disposed of. Crews would be sent overland to the Baltic Sea, and after the expenses of the trip were deducted, the proceeds were to be divided with one-fifth for the Russian government and the remainder to the captors, who could only be the state-controlled Russian-American company or officers of the Russian Navy. Such an arrangement naturally led to competitive enforcement of the ukase, and greatly relieved the state of the burden of policing its own laws. Parts of this ukase were recognized by the Anglo-Russian treaty of 1825 fixing the boundary between Russian and British possessions in North America, and have carried over into U.S.-Canada negotiations on the boundaries of Alaska (Green, 1982).

However, the principle of sovereign seas did not go unopposed. Evolving ideas of property and the rule of law, and the increased technical maritime competence of many nations, weakened and finally dispelled the notion of mare clausem as a general principle. In 1599 Queen Elizabeth I of England protested to the King of Denmark and Norway about the seizure of English fishing and trading ships in the North Sea, and for the first time invoked the "Law of Nations" in favour of the principle of open seas, or mare liberum. The Law of Nations, or international law, was slowly developing as a formulation of accepted codes of relations between nations. It received its major impetus when Hugo Grotius published Mare Liberum in 1609, a scholarly, reasoned analysis of the rights of nations to travel by sea to other nations, and to trade with them. His immediate purpose was to make a legal case to support Dutch trade with India against the Portugese decree; but in so doing he introduced the principle of the freedom of the seas, based on logical arguments that were consistent with those upheld by courts for jurisdiction and freedom on land. Despite masterful arguments such as those put forth by John Selden in 1635 in his book Mare Clausem commissioned by Kings James I and Charles I of England to prove that dominion over the sea belonged to the ruling power, the principle of freedom of the seas became entrenched in international law. Under this principle, the high seas should be open and free for the use of all nations, but each littoral state had jurisdiction over a strip of waters (territorial waters) near the coast and controllable from the land (Johnson Theutenberg, 1984 b). The principle of mare liberum was endorsed by most of the major world powers from the seventeenth to the twentieth century, with the exception of the Russian ukase in the North Pacific noted above; and it was encoded in the 1958 Law of the Sea declarations of the United Nations. Its evolution has left unattended those aspects of resources and control which were inherent in the notions of mare nostrum and mare clausem and which, in recent years, have given rise to issues surrounding Exclusive Economic Zones.

The Last Few Decades: A Time of Rapid Change

Today, we are witnessing something like another revolution of the Law of the Sea, perhaps as fundamental as those of the thirteenth and seventeenth centuries. Areas that for nearly 400 years have been accepted as "high seas", to which anyone could have access and in which all could ply their trades (including fishing), are now in one form or another coming under the sovereignty of coastal states. Three-quarters of the world's coastal states have within the past twenty-five years declared a national economic control zone extending beyond their previous territorial waters.

Principles other than international freedom and uncontrolled access now constrain large parts of previously open seas. For example:

- in 1945 - the United states proposed national jurisdiction over seabed resources on continental shelves outside territorial waters;
- in 1952 - The Santiago Declaration extended exclusive fishing rights to 200 miles from the coast line;
- in 1958 - The United Nations adopted the principle of national jurisdiction over resources on the seabed of continental shelves;
- in 1970 - Canada proclaimed anti-pollution jurisdiction to 100 miles in arctic waters;
- in 1977 - The North Sea ceased to be international waters either for fishing or seabed resources development, and all of it became "taxable territory".

There are many other examples, in virtually all parts of the world, of what Ambassador Johnson Theutenberg (1984 a) has called "creeping jurisdiction" over international waters in the past two decades.

The change is rapid, and profound. The reasons are complex. The consequences, and the future, are unclear.

Causes of the New Situation

What are the new factors that have caused the 400-year-old principle of mare liberum - which is still endorsed by most of the navies and shipping interests of the world - to become unsatisfactory to governments of most coastal states, or to be challenged in recent decades?

There are many reasons, some having little to do with the oceans. But the following are important to nearly every example:

1. Development of new technologies, from sonar fish-finders and freezer ships to offshore drilling platforms and weather satellites have on the one hand greatly expanded the possibilities for development of marine areas and opened up the world oceans to all who possess the technology and the capital to employ it, and on the other hand have left no area "safe" from the activities of others, under the old rules. These new technologies have led to the possibilities of new kinds of marine resources, whose exploitation may be at odds with traditional uses of the seas. For example, fifteen percent of the world's petroleum now comes from the oceans, and the percentage is increasing. Mining of seabed minerals, production of energy from ocean temperature gradients, enhanced breeding of free-roaming but genetically programmed fish stocks, are possibilities in the not too distant future. Many of these new developments are incompatible with one another or with traditional activities, and raise new problems of management and control. The new technologies have led to new kinds of economic activity, at national, international, and multinational scales. They have developed new marine physical entities, such as oil well platforms, artificial islands, undersea habitats, seabed pipelines that cross "national" jurisdictions, to which the old rules cannot easily be applied;

2. Depletion of living resources, through over-fishing (sometimes apparently through use of disruptive technologies more than through actual fish

catches, as in the dispersal of herring schools by netting from the middle of a school that has been located acoustically and leaving the schools vulnerable to natural predators), disruption of habitat, especially of the critical breeding areas for migratory fish and marine mammals, and disturbance of marine ecosystems. The hunting philosophy which impels men to move on to richer, more distant natural areas when nearby areas are depleted, rather than the cultivation impulse to manage near-by areas to increase yields on a sustained basis, is still dominant in fishing activities; and combined with modern ships and processing technologies, has made it possible for any modern nation to fish profitably in any part of the world where fish are still abundant. Thus we have Japanese trawlers in the North Atlantic, Soviet fishing in the Indian Ocean, and several nations, having depleted the whales in the Antarctic, turning to krill in the same area. It is no surprise that nearly every marine state now wants to protect its own nearby fish stocks but still have access to fish stocks elsewhere;

3. A new and integrated view of the role of oceans and coastal areas in the regional and global environment has made many of the formerly accepted principles of unrestricted use of marine commons unfeasible, or indefensible, on practical, scientific and political grounds.

Scientifically - or perhaps one should say philosophically - it has always been recognized that the oceans, the atmosphere and the land were intimately linked and the nature or characteristics of each depended on the others. The oldest geophysical treatise known, the Babylonian Enuma Elish written about 1500 B.C. begins with beautiful imagery of the hydrological cycle, with the wealth and order of the universe being born out of watery chaos and the struggle between sea, land, and atmosphere (Jacobsen, 1957). This detailed story of the Earth's beginnings, and in fact most early cosmologies, tell how disturbance of the ocean, usually expressed by battles between gods representing primitive elements, affects the climate and the nature or resources of the land. Such stories show how early people in nearly all societies were keen observers of the inter-relationship of environmental processes. Nevertheless, in more recent times we have tended to ignore the interdependence of ocean, land, and atmosphere, and our laws and practices assume that the ocean, in particular, is too big and stable to be vulnerable to environmental change, especially to change caused by human activities. Today, the weight of evidence is convincing that the oceans are indeed vulnerable to change, that marine environmental quality is a determining factor in the quality of the global and terrestrial environment, and that human activities must be managed to protect the environment of the oceans just as importantly as to protect the land and the quality of the lower atmosphere (Economic Summit, 1985). This change in perspective has important implications for national and international responsibilities toward the marine environment (IUCN, 1980).

The Oceans Today -- and in the Future?

The oceans, including coastal areas, can be seen to be serving five functions from the human point of view. Each of these is distinct in relation to human activities or needs, but they are inseparable in the environmental and natural response of the oceans themselves, and thus our use for one purpose affects other uses. The integration, by the oceans themselves, of the effects of our uses of them, is a major cause of the issues surrounding Exclusive Economic Zones. The five functions of the oceans are:

(i) they are a source of renewable resources. Their productivity and the sustainability of resource supply depends on the maintenance of favourable environmental conditions over areas much larger than those adjacent to any coastal state;

(ii) they are a medium of transportation and communication (sometimes indeed a barrier) whose utility depends upon agreed or enforceable principles of access and use (this includes of course, military operations);

(iii) they may be locations for development of non-renewable resources, whose successful exploitation depends upon a stable system of management and control, and adequate information on natural conditions;

(iv) they are the depository for most of mankind's persistent wastes, receiving them either directly from deliberate discharge or indirectly and inadvertantly through delivery by natural environmental processes of atmospheric and hydrological circulation;

(v) they play a major role in maintaining and regulating the world's climate and chemical environment and the productivity of planetary biological systems:- a role that may be more dynamic and less stable than earlier supposed.

The first two of these functions served the traditional uses of the seas. It is around them that legal and political concepts of mare clausem and mare liberum developed, and for several centuries they have been served fairly well by principles of mare liberum and territorial waters. The third and fourth functions have become of major importance only in recent years, without taking away from the importance of the first two, and the fifth, long taken for granted, is becoming seen to be of fundamental importance for all nations and all creatures and, to a new and sobering degree, a collective responsibility of Mankind.

All five "functions" are involved in the issues surrounding Exclusive Economic Zones. An imposed regime that neglects any one of them, in any but the smallest geographical area, is likely to be unworkable, politically, economically, or environmentally. It is these questions that have brought us here today; and to look at them, the conference organizers have divided our discussions into four sessions:

A - A review of the legal regime and the United Nations Convention on the Law of the Sea;

B - Questions of resource development and exclusive economic zones;

C - Resource assessment in marine areas;

D - Questions of policy and management.

References

Blom, G.A., 1984. "The Participation of the Kings in the Early Norwegian Sailing to Bjarmeland (Kola Peninsula and Russian Waters) and the Development of a Royal Policy Concerning the Northern Waters in the Middle Ages"; Arctic, vol. 37, No. 4, pp. 385-388.

Economic Summit, 1985. "Report on the Environment by the Technology, Growth and Employment Working Group" Report to the Bonn Economic Summit, 1985. Her Majesty's Stationery Office, London, ISBN 0-10-195000-4; Technical Report 3, pp. 44-50.

Fulton, T.W. 1911. "The Sovereignty of the Sea", Blackwood, Edinburgh and London, 480 pp.

Green, L., 1982. "The Boundary Hunters", University of British Columbia Press, 214 pp.

IUCN (International Union for the Conservation of Nature and Natural Resources) 1980. "The Global Commons"; in World Conservation Strategy, Report by IUCN, United Nations Environment Programme and World Wildlife Fund with the assistance of the Food and Agriculture Organization and UNESCO (2 vols.) Chapter 18.

Jacobsen, T., 1957, "Enuma Elish, 'The Babylonian Genesis'", in Theories of the Universe, M.K. Munitz, ed., Free Press, Glencoe Illinois, pp. 8-20.

Johnson Theutenberg, B., 1984(a). "The Evolution of the Law of the Sea"; Natural Resources and Environment Series, Vol. 17, Tycooly International Publishing Limited, Dublin, 261 pp.

Johnson Theutenberg, B., 1984(b). "Mare Clausem et Mare Liberum"; Arctic, vol. 37, No. 4, pp. 481-492.

EXCLUSIVE ECONOMIC ZONES

THE LEGAL REGIME AND THE UN CONVENTION ON THE LAW OF THE SEA

E.D. BROWN

Professor of International Law in the University of Wales; Director, Centre for Marine Law & Policy, UWIST, Cardiff

1. THE EEZ IN HISTORICAL PERSPECTIVE

At the end of the Second World War, the seas were still divided into only four jurisdictional zones: internal waters, lying landward of the baseline of the territorial sea; the territorial sea, the breadth of which was uncertain but generally considered not to exceed 12 miles at most; the contiguous zone of indeterminate breadth, claimed by only some States and for very limited jurisdictional purposes; and, finally, the high seas. Today, in 1986, five more zones are generally recognised or are in the course of attracting general recognition: the continental shelf, the exclusive fishing zone, the exclusive economic zone, archipelagic waters and the "Area" beyond the limits of national jurisdiction. It may, therefore, help to see the EEZ in historical perspective if a brief account is given of the principal landmarks of the period 1945-86.

The first, and arguably, the most important development was the assertion of jurisdiction and control over the natural resources of the continental shelf by President Truman in his celebrated Presidential Proclamation No. 2667 of 28 September 1945.[1] This, as the International Court of Justice (ICJ) recognised some 24 years later,[2] came to be acknowledged as the starting point of the positive law on the continental shelf. For the parties to it, that law was later codified and developed in the Geneva Convention on the Continental Shelf, 1958.

A second landmark was the introduction of the differentiation between the concept of the territorial sea and that of the exclusive fishing zone. Following the unsuccessful termination of UNCLOS II in 1960,[3] it was clear that it would no longer be possible to maintain one boundary marking the outer limit of both the coastal State's territorial sea and its exclusive fishing zone. The pace of developments thereafter was fast and furious. In a series of bilateral agreements between North Atlantic States, recognition was accorded to exclusive fishing zones beyond the territorial sea and in 1964 this trend was generalised in the European Fisheries Convention, which incorporated a six-plus-six formula: a six-mile territorial sea with an exclusive fishing zone in the further six-mile band beyond it.[4] More recently, following the trail blazed much earlier by a few Latin American States, Iceland, in a series of inflationary unilateral claims, pushed her fishing limit out first to 12 miles, then to 50 miles and, finally, to 200 miles.[5] After some initial hesitation, much of the rest of the world appears to have decided to follow suit and claims to a 200-mile exclusive fishing zone are now quite commonplace. Moreover, the ICJ has set the seal of its approval on the concept of the exclusive fishing zone, though without reference to any particular breadth.[6]

There can be little doubt that the trend set by Iceland was followed by other States because of the third landmark: the gradual consolidation of the new notion of an exclusive economic zone (EEZ) since it was introduced into the United Nations law of the sea debate in 1972.[7] By about 1975-76, it seems to have been generally accepted within the international community that the trend towards recognition of some kind of EEZ was now irreversible. The details might still be missing but the EEZ formula would certainly include recognition of the exclusive fishing rights of the coastal State in a zone extending to 200 miles from the baseline of the territorial sea. Part V of the UN Convention[8] now supplies the missing details in abundance, certainly at least for States which become parties to the Convention. More generally, in 1982 the ICJ referred to "the concept of the exclusive economic zone", even if only obiter, noting that it "may be regarded as part of modern international law"[9]; and more recently it expressed the view that it was "incontestable that ... the institution of the exclusive economic zone, with its rule on entitlement by reason of distance, is shown by the practice of States to have become a part of customary law."[10]

The general recognition of the new concept of archipelagic waters is of even more recent vintage. Although States such as Indonesia and the Philippines have been pressing the international community for some time to recognise their right to construct a system of straight baselines around the archipelagos, it was not until the UN Convention was adopted that the concept of archipelagic waters was accepted more widely.[11] To some extent, its general recognition is tied to the fate of the new Convention.

Finally, mention must be made of the "Area beyond the limits of national jurisdiction", that is, the area of the deep sea-bed lying seaward of the outer edge of the legal continental shelf. The conventional regime of the Area, developed by UNCLOS III and embodied in Part XI of the UN Convention and related Annexes and Conference Resolutions, gives effect to the radical notion that the Area and its resources are the common heritage of mankind.[12] As will be seen below, it is this exceedingly complex and controversial part of the Convention which is regarded by the United Kingdom Government and others as the main stumbling block in the way of accession to the UN Convention.

The maritime zones falling <u>within</u> the limits of national jurisdiction are represented in Figure 1, with the exception of internal waters and archipelagic waters, both of which lie landward of the territorial sea baseline.

Figure 1 helps to emphasise that some of these jurisdictional zones overlap. Thus, (i) the waters of the 24-mile contiguous zone also fall within the 200-mile EEZ; (ii) the sea-bed and subsoil of the EEZ coincide with those of the continental shelf out to the 200-mile limit; (iii) the high seas overlap that part of the continental shelf which extends beyond 200 miles.[13] This is another way of saying that the superjacent waters of the continental shelf (but not of the EEZ) remain high seas.

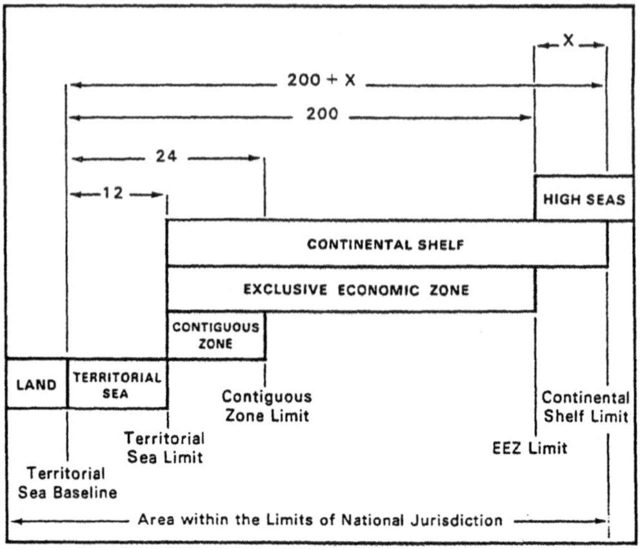

Figure 1. The area within the limits of national jurisdiction: jurisdictional zones

2. THE JURIDICAL NATURE OF THE EEZ

The concept of the EEZ has been developed both in State practice (unilateral declarations, statutes etc.) and in the Third United Nations Conference on the Law of the Sea (UNCLOS III). State practice has led to the establishment of the EEZ as a concept received into international customary law. UNCLOS III produced the UN Convention, Part V of which embodies a detailed conventional formulation of the notion of the EEZ. The concept is not necessarily the same under international customary law and the Convention. In this section the juridical nature of the zone is explored. This involves comparing the EEZ with other maritime zones, thus emphasising its unique sui generis nature, examining the content of typical national EEZ claims, questioning the accuracy of "Exclusive" in Exclusive Economic Zone and considering whether it is necessary to assert a claim to an EEZ in order to acquire rights in it.

2.1 The sui generis character of the EEZ

Prior to the introduction of the notion of the EEZ, it was true to say that the outer limit of the territorial sea marked a fundamental divide among offshore jurisdictional zones. Thus, landward of that line it was recognised that the predominant, fundamental principle was that of sovereignty. In the most landward area, that of the internal waters lying inside the territorial sea baselines, the sovereign powers of the coastal State were practically unlimited. Even in the belt of territorial sea, the only major limitation upon the coastal State's sovereignty was the duty to permit the innocent passage of foreign shipping. The essential point to note is that the residual status of these two zones was that of sovereignty. To put it another way, there were legal presumptions in favour of the sovereign powers of the coastal State which, in case of doubt, would override the interests of other States in these zones. Any rights of these States, such as innocent passage, were to be regarded as exceptions to the underlying principle and, as exceptions, to be restrictively interpreted.

Seaward of the outer limits of the territorial sea, the legal presumptions were reversed. Thus, in the event of conflict between the interests of the coastal State and those of the international community, the latter, founded on the principle of the freedom of the seas, would prevail. Prior to the emergence of the concept of the EEZ, three important zones were generally recognised in this area beyond the outer limit of the territorial sea, namely, the contiguous zone, the continental shelf and the exclusive fishery zone. All three accord specified, limited rights or powers to the coastal State and the essential point to note is that such rights and powers constitute exceptions to the underlying fundamental principle of the freedom of the high seas. For example, in the continental shelf, the coastal State enjoys sovereign rights for the purpose of exploring the continental shelf and exploiting its natural resources but such rights do not affect the legal status of the superjacent waters as high seas.

In case of doubt, conflicts between the continental shelf rights of the coastal State and the rights of other States to enjoy the various freedoms of the high seas in the superjacent waters would be resolved by reference to the legal presumptions flowing from the fundamental principle governing this area, namely the freedom of the high seas.

In the same way, the exclusive fishery rights of the coastal State in the exclusive fishing zone were regarded as exceptions to the underlying status of the area as high seas.

It is in this respect that the EEZ differs from the other maritime zones lying seaward of the territorial sea. In describing the status of the EEZ as being sui generis, the intention is to indicate that it departs from this common pattern. If the common pattern had been followed, the EEZ would have been an area of the high seas in which certain rights were accorded to the coastal State by way of exception to the fundamental principle of the freedom of the high seas. In fact, quite a different pattern is followed in Part V of the UN Convention and its main characteristics may be summarised as follows:

(i) the EEZ area is no longer regarded as part of the high seas but is defined in Article 55 as being:

> ... an area beyond and adjacent to the territorial sea subject to the specific legal regime established in this part, under which the rights and jurisdiction of the coastal State and the rights and freedoms of other States are governed by the relevant provisions of this Convention.

The absence of any reference to the high seas in this definition, taken with the words underlined, clearly signals the departure from the earlier common pattern.[14]

(ii) Article 56 specifies the rights, jurisdiction and duties of the coastal State in the EEZ.

(iii) Article 58 specifies the rights and duties of other States in the EEZ.

Now, under the old pattern, conflict between coastal States' rights and other States' rights would have been resolved in case of doubt by reference to the freedom of the high seas and if particular rights had not been accorded to the coastal State by the Convention, it would have been presumed that the coastal State did not have those rights. However, Article 59 of the UN Convention embodies quite a different approach by providing that:

> In cases where this Convention does not attribute rights or jurisdiction to the coastal State or to other States within the exclusive economic zone, and a conflict arises between the interests of the coastal State and any other State or States, the conflict should be resolved on the basis of equity and in the light of all the relevant circumstances, taking into account the respective importance of the interests involved to the parties as well as to the international community as a whole.

In short, there are no longer any legal presumptions concerning rights and jurisdiction in the EEZ.

2.2 National EEZ claims

According to a United States government report published in 1985,[15] EEZ claims have been made by 64 countries, though it is understood that a forthcoming book reprints the claims of 70 countries.[16] The earliest claims to EEZs as such (excluding 200-mile territorial sea claims and exclusive fishery claims) appear to have been made in 1974 but the great bulk of them have been made since 1977, the year in which UNCLOS III produced the Informal Composite Negotiating Text.[17] Although drafting changes were subsequently made to the EEZ provisions of that text, it is true to say that in their essentials they had assumed their final form in that 1977 text. This is of course significant because much of the national legislation passed at about that time or subsequently follows closely the substance and in many cases the form of that text. However, it is also true that there are a great many variations on the theme, partly because some States modelled their legislation on earlier negotiating texts which, for example, accorded "jurisdiction" or "exclusive jurisdiction" over the production of energy from the water, currents and winds, rather than the "sovereign rights" granted by Article 56(1)(a) of the final UN Convention.[18]

The variety displayed by national legislation and, in many cases, the lack of detailed provisions as compared with Part V of the UN Convention, make it difficult to identify with precision the concept of the EEZ as developed in this State practice or to say how it differs, if at all, from the concept as formulated in the UN Convention. It is probably fair to say, however, that as time goes on and more and more States model their legislation on the UN Convention, its influence will become ever more predominant and render insignificant any variations in the earlier legislation.

2.3 An Exclusive Economic Zone?

Given what has already been said about the sui generis nature of the EEZ - that it is something different from both the territorial sea and the high seas - it will be obvious that the coastal State does not have sovereignty over this zone. It may be less obvious, but it is nonetheless the case, that the sovereign rights of the coastal State over the economic resources of the zone are by no means as exclusive as the name of the zone might suggest. For example, as will be seen below,[19] the coastal State's sovereign rights over EEZ fisheries have to be exercised subject to clear obligations to grant access to these resources to other States in specified circumstances.[20]

2.4 The need to claim an EEZ

Under both the Geneva Convention on the Continental Shelf, 1958 and the UN Convention, the coastal State has no need to claim a continental shelf; it enjoys its entitlement ipso jure, by automatic operation of the law. There is no corresponding provision in the UN Convention so far as the EEZ is concerned and it is generally recognised that a State must claim an EEZ in order to be entitled to it. Given the fact that, under the UN Convention, coastal States are entitled to a continental shelf of at least 200 miles in breadth (where geographically feasible), the combination of these two rules produces a curious result. It means that States have an automatic entitlement to the bed and subsoil of the 200-mile EEZ, which coincides with the continental shelf, but have to claim the superjacent water column. It may be noted in passing that, under Article 56(3) of the UN Convention, the coastal state's rights in the sea-bed and subsoil of the EEZ are to be exercised in accordance with Part VI, that is, the Part dealing with the continental shelf.

3. THE SUBSTANCE OF PART V OF THE UN CONVENTION

3.1 Rights and duties of the coastal State and of other States in the EEZ: framework provisions in Articles 56 and 58

Articles 56 and 58 are framework provisions in the sense that they state in general terms the rights, jurisdiction and duties of the coastal State and the rights and duties of other States, leaving the details to be filled in by later articles.

3.1.1 Rights and duties of the coastal State - Article 56

Article 56(1) deals in turn with (a) the coastal State's "sovereign rights", (b) coastal State "jurisdiction" and (c) "other rights" provided for in the Convention. Comment will be confined to (a) and (b).

(a) Sovereign rights. Paraphrasing the provisions of Article 56(1)(a), the coastal State has sovereign rights over (1) living resources in the EEZ water column; (2) non-living resources in the water column and sea-bed and subsoil; and (3) other activities for the economic exploitation and exploration of the zone, such as the production of energy from the water, currents and winds.

What is meant by "sovereign rights"? Perhaps it can be said that while "sovereign rights" are less than full sovereignty over the area, they are more than simply exclusive rights, in the sense that their scope must be extensively interpreted to include whatever powers are needed to ensure that the coastal State is enabled to enjoy its rights in full. However, as noted above, these rights are by no means absolutely exclusive since their exercise is limited in certain cases by the obligation to share with other States.

It is perhaps necessary to emphasise the reference to the coastal State's having sovereign rights "with regard to other activities for the economic exploitation and exploration of the zone", a clause which in effect extends the coastal State's sovereign rights beyond EEZ fisheries and continental shelf resources to embrace all economic activities in the zone.

(b) Jurisdiction. The concept of "jurisdiction" is of a lower order than "sovereign rights" and, indeed, when unqualified by any such adjectives as "exclusive" or "absolute" is rather vague in its scope. Under Article 56(1)(b), the coastal State has jurisdiction with regard to (i) the establishment and use of artificial islands, installations and structures; (ii) marine scientific research; and (iii) the protection and preservation of the marine environment. In all these cases, the scope of the coastal State's jurisdiction has to be determined by reference to "the relevant provisions of" the Convention, some of which will be examined below.

The "due regard" rule. That the rights of the coastal State in the EEZ are not absolute rights is emphasised again by the stipulation in Article 56(2) that, in exercising its rights and performing its duties under the Convention, the coastal State must have due regard to the rights and duties of other States. The substance of these rights and duties of other States must now be considered.

3.1.2 Rights and duties of other States - Article 58

Much of the opposition to the introduction of the new concept of the EEZ, expressed in the course of UNCLOS III, was motivated by fears that it would put in jeopardy the enjoyment by the international community of the traditional freedoms of the seas hitherto applicable beyond the outer limit of the territorial sea. It is the purpose of Article 58 to assuage these fears and embody a balance between the rights of the coastal State and those of the wider international community. It does so by preserving some of the classical freedoms which are imported into Part V of the Convention by reference to the provisions of Part VII on the High Seas.

Of course, not all of the high seas freedoms are preserved in the EEZ. Article 58(1) refers only to what might be called the freedoms of transport and communication. Thus the freedoms of navigation and overflight and of the laying of submarine cables and pipelines are specifically mentioned, together with "other internationally lawful uses of the sea related to these freedoms, such as those associated with the operation of ships, aircraft and submarine cables and pipelines, and compatible with the other provisions of [the] Convention".

There are two points to notice about these provisions. First, even these freedoms are to be enjoyed "subject to the relevant provisions of this Convention". For example, under Article 79(3), the delineation of the course for the laying of pipelines is subject to the consent of the coastal State. Secondly, by specifying certain freedoms and ignoring others, this Article underlines the point, mentioned above, that there is no longer any presumption in favour of other freedoms.

It is of course obvious that several of the high seas freedoms listed in Article 87 (in Part VII on the High Seas) could no longer survive in the EEZ. As has been seen, three of these freedoms have been transformed into the coastal State's sovereign rights over fishing, and its jurisdiction over artificial islands etc. and scientific research. However, this leaves other freedoms, such as the freedom to conduct naval exercises, which, though not referred to specifically in Article 87(1), would certainly fall within its preambular "inter alia" formula. Unless one were prepared to say that naval exercises are simply a species of navigation, it would have to be recognised that they fall into the category catered for by Article 59 - rights or jurisdiction not attributed to either the coastal State or to other States. Accordingly, any conflict over State A's naval exercises in State B's EEZ would have to be resolved "on the basis of equity and in the light of all the relevant circumstances, taking into account the respective importance of the interests involved to the parties as well as to the international community as a whole." That there will be controversy over the subject is already clear from the interpretative declarations made by a number of States when signing or ratifying the Convention.[21] Brazil, Cape Verde and Uruguay took the position that military exercises could be conducted in the EEZ only with the consent of the coastal State. Italy, on the other hand, contended that the rights and jurisdiction of the coastal State "do not include the right to obtain notification of military exercises or manoeuvres or to authorise them."[22]

The applications of high seas rules in the EEZ. Under Article 58(2), the whole corpus of rules relating to the high seas, contained in Articles 88 to 115 of the Convention are made applicable to the EEZ, though subject to the condition that they are compatible with Part V on the EEZ. It follows, for example, that the rules on piracy, which at first sight appear to be confined to the high seas, defined as excluding the EEZ, are in fact applicable in the EEZ.

The "due regard" rule. It was noted above that Article 56(2) imposes a "due regard" duty upon the coastal State. Article 58(3) reciprocates by imposing a like duty upon other States. It also requires them to comply with the laws of the coastal State in the EEZ in so far as they are not incompatible with Part V.

3.2 The EEZ living resources regime

This is a very large topic a comprehensive discussion of which is beyond the scope of this paper. It may, however, be useful to comment on two aspects of the fishery regime. The first is the degree to which the coastal State's sovereign rights are limited by the terms of the Convention; and the second is the related question of the provision made for settlement of disputes.

3.2.1. The scope of the coastal State's sovereign rights

It was noted above that the coastal State's sovereign rights are limited by its obligation to share the living resources of the EEZ with other States in certain circumstances. One of the questions which has constantly to be asked about the EEZ living resources regime is how real and enforceable are these obligations to share. It is certainly arguable that the restraints on the exercise of the sovereign rights of the coastal State are minimal, especially in view of the weakness of the provisions on dispute settlement. The proposition may be illustrated by reference to the provisions of the Convention on conservation

(Article 61), utilization (Article 62) and dispute settlement (Article 297).

So far as conservation is concerned, the provisions of Article 61 may be summarised as follows:

(1) The coastal State determines the allowable catch.

(2) There is an obligation upon the coastal State to ensure that "the maintenance of the living resources ... is not endangered by over-exploitation.

(3) The coastal State's conservation and management measures are to be designed also to maintain or restore populations of harvested species at levels which can produce the maximum sustainable yield (MSY), as qualified by relevant environmental and economic factors, including the economic needs of coastal fishing communities and the special requirements of developing States, and taking into account fishing patterns, the interdependence of stocks and any generally recommended subregional, regional or global minimum standards.

It may be noted in passing that this formula and especially the criteria by reference to which the MSY may be qualified are extraordinarily wide and vague and there is a potential conflict between the coastal State's sovereign rights and its obligation to apply the heavily qualified MSY formula. In this situation, the coastal State determines in the first instance what measures are required to achieve such a qualified MSY, only "taking into account" scientific evidence and, "as appropriate", co-operating with international organisations. What happens when there is a dispute over such measures will be considered later.

(4) The effects of the conservation measures on associated species have to be taken into consideration.

(5) Provision is made for exchange of scientific information.

Turning next to the provisions of Article 62 on utilization of living resources:

(1) The coastal State has to promote the objective of <u>optimum utilization</u> without prejudice to Article 61. In other words, under Article 61(1) the coastal State establishes an allowable catch which, under Articles 61(2) and (3) must mean that which will allow maintenance of a heavily qualified MSY. Article 62(1) is thus concerned with the optimum utilization of this total allowable catch (TAC).

(2) The coastal state determines its own harvesting capacity (HC) and if HC is less than TAC, it must grant access to the fishermen of other States to the surplus. Such access has to be in accordance with agreements or other arrangements and pursuant to the terms, conditions and regulations referred to in Paragraph 4, having particular regard to the provisions of Articles 69 and 70 (which deal with the rights of land-locked and geographically disadvantaged States), especially in relation to developing States.

(3) In giving access to the surplus catch, the coastal State must <u>take into account</u> all relevant factors, including (i) the economy of the coastal State; (ii) its other national interests; (iii) Articles 69 and 70; (iv) the requirements of the developing States of the region or subregion; and (v) the need to minimise economic dislocation in States which habitually fished in the zone or had made substantial efforts in research and identification of stocks. But, of course, the degree to which such "taking into account" might influence the final carve-up of the TAC would once again depend upon the decisions of the coastal State in the first instance.

(4) There is an obligation upon nationals of other States to comply with the conservation measures and with other terms and conditions established in the regulations of the coastal State. It is true that these regulations have to be consistent with the Convention, but the coastal State still has a wide discretion in formulating such regulations on, <u>inter alia</u>, licences and fees; species to be caught and quotas; fishing seasons, areas, gear and vessels; age and

size of fish; landing of catch in coastal State; observers and trainees on board; and enforcement procedures.

The above summary of the provisions of Articles 61 and 62 amply demonstrates that the Convention gives the coastal State not only the paramount role in applying the very broad and vague rules on conservation and utilization of fish stocks but leaves to it very wide discretions at least in the first instance. Moreover, if time permitted a similar review of Articles 69 and 70, it would be seen that they too combine vague formulae with extensive coastal State discretion in the first instance. It is of course true that such provisions may be entirely satisfactory if the will is there to apply them fully and fairly. The more important question, however, is to discover what is likely to happen if that will is lacking and disputes arise over the failure to apply these provisions.

3.2.2 Dispute settlement

Disputes arising under the EEZ fisheries regime are subject to Sections 1 and 2 of Part XV of the Convention, which relate to "General provision and General Obligations" and "Compulsory Procedures Entailing Binding Decisions" respectively. However, under Section 3, Article 297(3), the coastal State is not obliged to accept the submission to such binding settlement of any dispute relating to its sovereign rights with respect to the living resources of the EEZ or their exercise. This escape clause covers, inter alia; the coastal State's discretionary powers for determining the allowable catch, its harvesting capacity, the allocation of surpluses to other States and the terms and conditions established in its conservation and management regulations. The most onerous obligation which the coastal State has to accept in relation to such disputes is that of submission to compulsory conciliation. However, this is compulsory only in the sense that the coastal State is bound to submit the dispute to the conciliation commission; the report of the commission is not binding and, in any event, under Article 297(3)(c), in no case may the conciliation commission substitute its discretion for that of the coastal State.

Conclusion. Given the combination of loosely drafted rules, the degree of discretion which is left to the coastal State and the obvious weakness of the dispute settlement procedure, it must be concluded that the EEZ living resources regime in the Convention imposes few restraints upon coastal States which may wish to transform their limited sovereign rights into practically unlimited sovereignty over the living resources of the zone.

3.3 Artificial islands, installations and structures

It was noted above that under the framework provision of Article 56(1)(b)(i), the coastal State has jurisdiction as provided for in the relevant provisions of the Convention with regard to the establishment and use of artificial islands, installations and structures. The principal "relevant provisions" are to be found in Article 60, Paragraph 1 of which gives the coastal state the exclusive right to construct and to authorise and regulate the construction, operation and use of three categories of offshore entities.

Dealing first with "installations and structures for the purposes provided for in article 56 and other economic purposes", it is quite clear that this category is confined to entities which have an economic purpose, though it may not always be easy to determine whether a particular activity is one with an economic as distinct from a pure science or other purpose.

The second category - that of "artificial islands" - is, however, referred to in quite unqualified terms and raises the question whether artificial islands too are permitted only if they are for economic purposes. The obvious *a contrario* argument suggests that no such limitation is intended, while the fact that they are provided for in the new regime of the exclusive *economic* zone suggests that such a limitation is implied.

The third category comprises "installations and structures which may interfere with the exercise of the rights of the coastal State in the zone". This formulation too is ambiguous. If such installations and structures are simply a species of the broader category of installations and structures with an economic purpose, why did they need separate mention? If they are not necessarily included within that broader category, for what purposes may they be established? Already, some States have filed interpretative declarations on this point. Brazil has interpreted the Convention as requiring the coastal State's consent whatever "the nature or purpose" of the installations or structures.[23] Cape Verde and Uruguay adopt a similar approach, requiring their consent for installations and stuctures be they those referred to in the Convention or any other kind.[24] Italy, on the other hand, believes that the coastal State's jurisdiction is limited to "the categories of such installations and structures as listed in article 60 ...",[25] by which they presumably mean those established for one of the purposes envisaged in Article 56.

Reference must be made to Article 60(2)-(8) for the more detailed rules concerning the construction, operation and decommissioning of such installations and structures.

3.4 Marine scientific research and the protection of the marine environment

It was noted above that, under Article 56, the coastal State has jurisdiction with regard to these subjects "as provided in the relevant provisions" of the Convention. It is beyond the scope of this paper to consider these aspects and it must suffice to draw attention to Part XII on the environmental question and Part XIII on marine scientific research.

4. CURRENT STATUS OF CONVENTION AND UK POSITION

Although, as at 3 February 1986, the UN Convention had been signed by 155 States and 4 other entities, and had attracted 26 ratifications, it still fell far short of the target of 60 ratifications or accessions required for entry into force under Article 308. It is notable too that the list of signatories and ratifying States does not include the Federal Republic of Germany, the United Kingdom or the United States.

The position of Her Majesty's Government has been made quite clear.[26] The barrier to accession remains the provisions of Part XI of the Convention, embodying the regime of deep sea mining, described by a Foreing Office Minister as "fundamentally unacceptable and unattractive to the Government."[27] The Government continues to play an active part in the Preparatory Commission, which is engaged in the task of developing the detailed regulations required for the entry into operation of the common heritage regime of sea-bed mining.[28] However, there are no indications that this work will render that regime any more attractive to the Government, especially as Germany and the United States also find it unacceptable. Whether the position will change with changes of Government in these States remains to be seen. In the meantime, there is no need for despondency so far as the United Kingdom's EEZ entitlement is concerned. The United Kingdom already has sovereign rights over continental shelf resources and has had an unchallenged 200-mile exclusive fishing zone since 1977. Moreover, as was noted earlier, the EEZ is now a concept received into the corpus of international customary law. While none would deny the desirability of having the advantage of the more precisely formulated rules in Part V of the Convention, their absence is not a significant disadvantage.

FOOTNOTES

1. United Nations, Laws and Regulations on the Regime of the High Seas, Vol. 1 (United Nations Legislative Series, ST/LEG/SER.B/1, 11 January 1951), pp. 38-40.
2. North Sea Continental Shelf, Judgment, I.C.J. Reports 1969, p. 3, at pp. 32-33, para. 47.
3. The very meagre results of UNCLOS II are reflected in its Final Act, Document A/CONF.19/L.15 (UNCLOS II Off. Rec. Annexes and Final Act, A/CONF.19/8, pp. 175-176).
4. See further E.D. Brown, "British Fisheries and the Common Market", 25 Current Legal Problems (1972), pp. 37-73.
5. See further, E.D. Brown, "Iceland's Fishery Limits: the Legal Aspects", 29 The World Today (1973), pp. 68-80.
6. Fisheries Jurisdiction (United Kingdom v. Iceland), Merits, Judgment, I.C.J. Reports 1974, p. 3, at p. 23, para. 52.
7. On the origins of this concept, see E.D. Brown, "Maritime Zones: A Survey of Claims", in R. Churchill et al. (eds.), New Directions in the Law of the Sea, Vol. III (1973), especially at pp. 167-186.
8. The official text of the UN Convention on the Law of the Sea 1982 and the Final Act of UNCLOS III are published in United Nations, The Law of the Sea (UN, New York, 1983, Sales No. E.83.V.5). This publication also includes a useful brief account of the history of the Convention (at pp. xix-xxviii). Also available from HMSO as Misc. No. 11 (1983), Cmnd. 8941.
9. Continental Shelf (Tunisia/Libyan Arab Jamahiriya), Judgment, I.C.J. Reports 1982, p. 18 at p. 74, para. 100. Cf. President Reagan's Proclamation 5030, 10 March 1983 (XXII ILM (1983), p. 465), which includes this paragraph: "Whereas international law recognises that, in a zone beyond its territory and adjacent to its territorial sea, known as the Exclusive Economic Zone, a coastal State may assert certain sovereign rights over natural resources and related jurisdiction".

10. Continental Shelf (Libyan Arab Jamahiriya/Malta), Judgment, I.C.J. Reports 1985, p. 13, at p. 33.
11. See further E.D. Brown, Passage through the Territorial Sea, Straits Used for Navigation and Archipelagos (David Davies Memorial Institute of International Studies, 1974), Chap. 3.
12. On the regime of sea-bed mining, see further E.D. Brown, Sea-Bed Energy and Mineral Resources and the Law of the Sea, Vol. 2: The Area Beyond the Limits of National Jurisdiction, Graham & Trotman, London, 1986.
13. Under Article 76(1) of the UN Convention, the continental shelf extends out to at least 200 miles from territorial sea baselines. Where, however, the continental margin extends beyond this 200-mile line, the outer edge of the continental margin is taken as marking the outer limit of the continental shelf. In Fig. 1, that part of the continental shelf which may lie beyond 200 miles is indicated by "X".
14. See also Article 86 which applies Part VII ("High Seas") to "all parts of the sea that are not included in [inter alia] the exclusive economic zone".
15. US Department of State, Bureau of Intelligence and Research, Limits in the Seas, No. 36: National Claims to Maritime Jurisdiction, 5th Revision, 6 March 1985, at p. 10.
16. R. Smith, Exclusive Economic Zone Claims. An Analysis and Primary Documents, Kluwer Academic Publishers, 1986.
17. A/CONF.62/WP.10, 15 July 1977.
18. For a useful tabulation of key excerpts from national legislation on the EEZ, see Law of the Sea Bulletin, No. 2 March 1985 (Office of the Special Representative of the UN Secretary-General for the Law of the Sea), pp. 1-96.
19. See below, section 3.2.
20. Similarly, the coastal State's jurisdiction over marine scientific research is not absolute. See further Part XIII of the UN Convention.
21. Law of the Sea Bulletin, No. 5, July 1985, at p. 45.
22. Ibid.
23. Ibid., at p. 46.

24. *Ibid.*

25. *Ibid.*

26. For a more comprehensive review of the British Government's position, see E.D. Brown, "The United Nations Convention on the Law of the Sea 1982: The British Government's Dilemma", 37 *Current Legal Problems* (1984), pp. 259-293.

27. See further *ibid.*, at p. 277.

28. See further Brown, *op. cit.* in note 12, Chapter 7.

MINERAL RESOURCES IN EXCLUSIVE ECONOMIC ZONES

D.S. Cronan
Marine Mineral Resources Programme
Imperial College, London SW7 2BP

INTRODUCTION

Imperial College launched a Marine Mineral Resources Programme two years ago in an attempt to draw together various aspects of work in this field which were then underway in the College, some of which had been active for a number of years, and to develop new research areas. Since its inception, the Programme has concentrated on mineral resource investigations in Exclusive Economic Zones. It is proposed, therefore, in this presentation to review the mineral resources known or suspected to be present in Exclusive Economic Zones, drawing on examples from Imperial College's Marine Mineral Resources Programme where appropriate.

There are four main groups of minerals on the sea floor, all of which occur to a greater or lesser extent within the Exclusive Economic Zones, claimed or projected, of some or other of the nations of the World. Two of these minerals, placers and phosphorites, occur at continental margins, placers generally in shallow water nearshore areas or on beaches, and phosphorites generally further offshore. Both are dominantly EEZ deposits. The other two, ferromanganese oxide deposits and metalliferous muds (including polymetallic sulphides), occur mainly on the deep sea floor. These only occur within EEZ's when the continental shelf is very narrow, allowing deep ocean conditions to intrude to within 200 nautical miles of the coast, such as occurs off some continental margins and around oceanic islands.

PLACERS

Placer minerals are deposits of generally heavy minerals that have been eroded from rocks on land and transported by rivers onto beaches and the near coastal offshore area. There are three main environments in which such deposits can occur (Fig. 1). The first of these is on beaches, either modern beaches, or ancient raised or submerged beaches. Black sands, for example, consisting largely of iron and titanium minerals, are a common occurrence on exposed tropical (non-coral fringed) beaches, and their deposition represents a balance between supply of the minerals from the land and the levels of wave energy required to concentrate them on the beaches. Conditions leading to placer mineral formation on modern beaches have been duplicated in the past, particularly during lower stands of sea level during the glacial periods. The beaches on which these placer minerals were formed are now submerged. During submergence of a placer mineral bearing beach, one of two things can happen. Either the encroaching sea can redistribute the placer minerals to form a disseminated deposit of wider extent and lower grade than the original deposit, or it can transport the minerals shorewards as it encroaches on the land to finally deposit them on or near the modern beach.

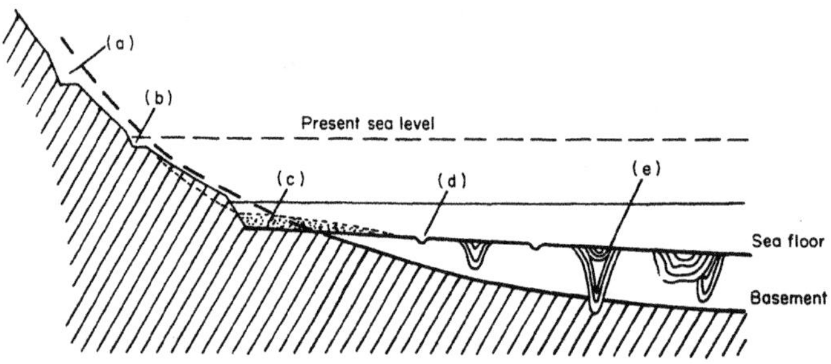

Fig. 1. Environments of possible placer mineral occurrence: (a) placers on raised beaches; (b) placers on modern beaches; (c) placers on submerged beaches; (d) placers trapped in surface depressions on the sea floor; (e) placers in buried river valleys. (From D.S. Cronan, 1980, Underwater Minerals, Academic Press; reproduced with the permission of Academic Press).

The second environment in which marine placer minerals can be found on the modern sea floor is in areas where a combination of wave and/or tidal conditions lead to the deposition of material being transported over the sea floor. The locations of these are very sensitive to the wave/tidal energy regime, and cannot easily be predicted.

The third environment in which placer minerals occur is in drowned river valleys. These represent river valleys which, during periods of lowered sea level, drained across now submerged areas and in which placer minerals were deposited by river action. The placer minerals have subsequently been buried by marine deposits.

World wide, placer minerals are found in both tropical, temperate and high latitudes. Major deposits of cassiterite (tin oxide) occur off Thailand and Malaysia, and minor deposits off Cornwall. Noble metal placers such as gold and platinum, occur off Alaska, and diamonds occur off S.W. Africa.

However, the most abundant placer minerals are the so called light heavy minerals, with specific gravity between about 3 and 5, such as rutile (Ti), ilmenite (Ti), chromite (Cr), zircon (Zr) and magnetite (Fe). These are widely distributed on high energy beaches in mid and low latitudes, and occur offshore of at least some of these beaches.

One area of light heavy minerals occurrence that has been investigated under Imperial College's Marine Mineral Resources Programme is in the Northern Aegean off Greece. This work is carried out in collaboration with the Greek Institute for Geology and Mineral Exploration. Light heavy minerals have been washed into the northern Aegean and are being redistributed by current action in the coastal zone. Statistical analysis has been carried out on geochemical data obtained on more than 500 surface sediment samples, which has indicated a coincidence of titanium and chromium bearing minerals in parts of the area. Zirconium appears to behave more independently. The statistical analyses, especially trend surface analysis, have been shown to be more reliable in picking out areas of heavy mineral concentrations than bulk element concentrations, and the chemical analyses are less time consuming than heavy mineral determinations which, anyway, only reflect the composition of the coarse fraction of the sediments.

PHOSPHORITES

Phosphorite deposits are composed mainly of calcium phosphate minerals of use in agriculture as a fertilizer. Most major marine phosphorites occur off the western margins of the continents where upwelling brings nutrients to the surface which support high biological productivity, but some occur off eastern continental margins (Fig. 2). The organic remains undergo decay on the seafloor thereby liberating phosphorus into the interstitial waters of the sediments which builds up to precipitate as phosphorites. Upwelling is particularly intense off the western margins of the continents, hence the preferential concentration of the phosphorites there.

Although quantitatively only of minor importance, phosphorites also occur on seamounts (Fig. 2) in oceanic island areas and on some of the islands themselves. Good examples are to be found in the equatorial Pacific and Indian Ocean. Although small in a global sense, these deposits could be of considerable importance to the economies of the island nations of the region, which are very agriculturally dependent. Valuable foreign exchange could be saved by their exploitation. One of the research topics being carried out under Imperial College's Marine Mineral Resources Programme

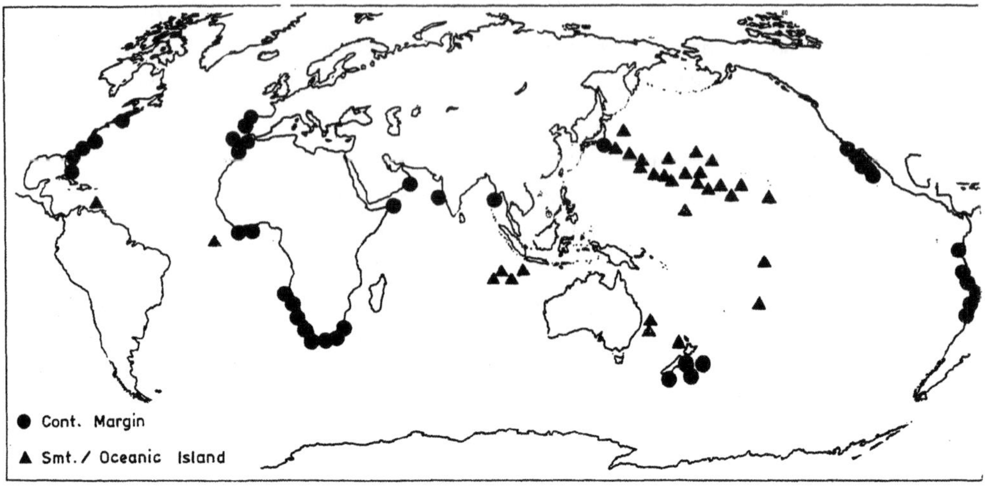

Fig. 2. Phosphorite locations on the sea floor. (From D.S. Cronan, 1980, Underwater Minerals, Academic Press; reproduced with the permission of Academic Press).

addresses exploration for these deposits in the equatorial western Pacific, using a computer based statistical analysis of multi-element geochemical data on S.W. Pacific sediments. The principle of the technique is to determine on the basis of inter-element associations in a multi-element data set of chemical analyses of sediments what proportion of the variability of a particular element, in this case phosphorus, can be explained by its co-variances with other elements. That part of it not explicable in this way, if it exceeds certain threshold values, can be regarded as anomalous, and probably reflects the presence of phosphorite deposits in the vicinity. A number of phosphorus anomalies have been found using this technique, many of them in sediments from on or near seamounts, which is the favoured location for phosphorite deposits on the S.W. Pacific sea floor. More detailed studies on these seamounts should confirm or deny the presence of phosphorites on them.

One very well studied phosphorite deposit occurs on the Chatham Rise, off New Zealand. Here the phosphorite lies on the sea floor within the Exclusive Economic Zone of New Zealand, and experiments on it have shown that it can be applied directly to New Zealand soils without prior treatment. This makes it attractive in an economy so heavily agriculturally dependent as that of New Zealand, and the Chatham Rise phosphorite may be brought into production before the end of the present century.

METALLIFEROUS SEDIMENTS AND POLYMETALLIC SULPHIDES

Metalliferous sediments and polymetallic sulphides consist of oxides of iron and manganese and sulphides of lead and zinc, and form by submarine hydrothermal processes associated with sea floor volcanism, mainly on mid-ocean ridges (Fig. 3). Seawater enters cracks and fissures in newly formed volcanic rocks on the sea floor, leaches metals from the rocks, and becomes transformed into a mineralising solution. On rising back to the sea floor under the influence of convection cycles, the minerals precipitate from it either within the rocks through which the solutions pass, or on their discharge into seawater. The latter situation can result in the formation of 'black smokers', which are plumes of finely divided sulphide precipitates. The 'black smokers' discharge through chimneys which can reach heights of 10 metres before collapsing to form a mound of broken up chimney material on which new chimneys grow (Fig. 4). The build up of 'smoker deposits' in this way can lead to quite large deposits of sulphide minerals, analagous to the so called massive sulphide deposits which are mined on land.

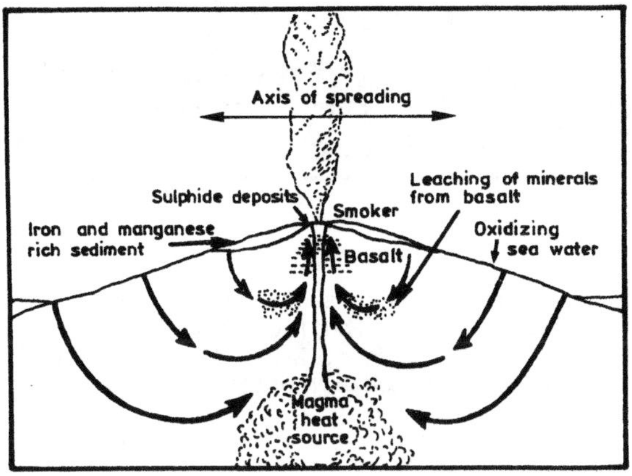

Fig. 3. Mid-ocean ridge smokers, sulphide deposits and associated manganese and iron rich sediments. (From D.S. Cronan, 1985, A wealth of sea floor minerals, New Scientist, June 6th; reproduced with the permission of New Scientist).

Metalliferous sediments and polymetallic sulphides were first studied in detail in the Red Sea where, in the Atlantis II Deep, they reach their greatest known development. Here all the known examples of sea floor hydrothermal deposits are found associated in a fractionation sequence comprising

Cu, Zn and Fe sulphides, Fe silicates and Fe and Mn oxides. Surrounding the deposits is a halo of manganese enriched sediments up to 20 km in diameter, the presence of which has served as the basis of a geochemical exploration model for the deposits which will be outlined below.

Although they have the potential to occur in any areas of mid-ocean ridges, principal interest in mid-ocean ridge metalliferous sediments and polymetallic sulphides is within EEZ's, especially off the west coast of North America. There a number of the deposits occur, both oxides and sulphides, in areas claimed by Mexico, the United States and Canada.

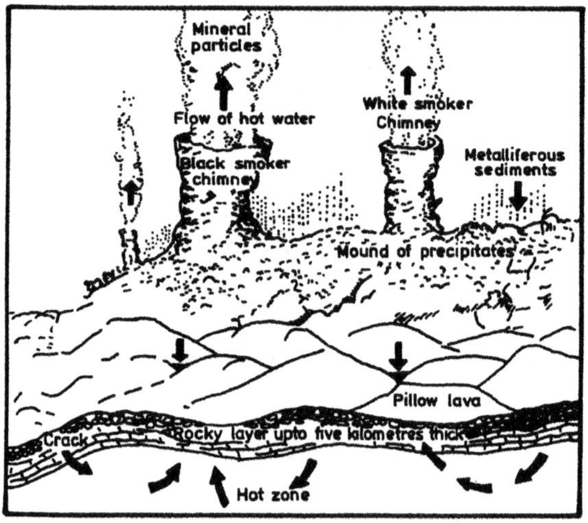

Fig. 4. A smoker mound: black smokers discharge sulphides and white smokers discharge silica (Modified from D.S. Cronan, 1985, A wealth of sea floor minerals, New Scientist, June 6th; reproduced with the permission of New Scientist).

After mid-ocean ridges, the second most important environment in which metalliferous sediments and polymetallic sulphides have been found is in island arcs (Fig. 5). These areas contain volcanic activity associated with subduction of oceanic plates, and have a potential to contain the hydrothermal deposits both on the arcs themselves and in "back arc" basins behind them. Their mechanism of formation is similar to that on mid-ocean ridges. One of the projects in Imperial College's Marine Mineral Resources Programme has been aimed at developing methods for exploring for such deposits in the S.W. Pacific island arcs. Initially, over 300 sediment samples were obtained from existing collections and were analysed for a variety of elements, including manganese. The results were plotted on maps of the region and, using the widespread distribution of manganese around hydrothermal centres discovered in the Red Sea and mentioned above, areas of likely hydrothermal deposits were outlined. These were then investigated in more detail and have revealed the presence of hydrothermal manganese deposits on the Kermadec Ridge, and polymetallic sulphide deposits in the Lau Basin and Bismarck Sea. These deposits fall within the EEZ's of the island nations of the region and thus are of considerable economic interest. Further exploration for such deposits in the region is underway.

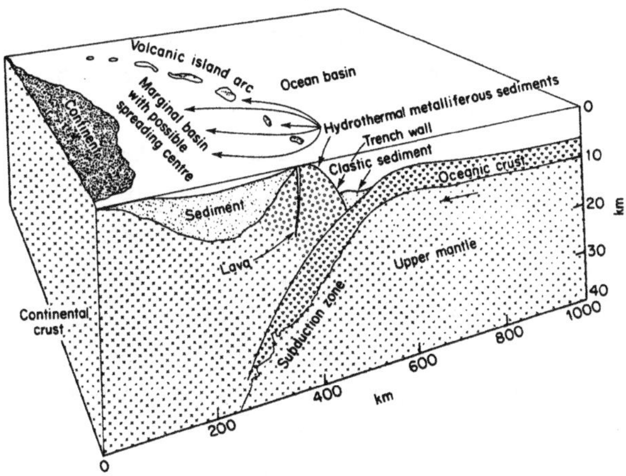

Fig. 5. Island arc and back arc basin with locations of metalliferous sediments. (From D.S. Cronan, 1980, Underwater Minerals, Academic Press; reproduced with the permission of Academic Press).

MANGANESE NODULES AND CRUSTS

Manganese nodules and crusts occur in all the oceans of the world, but are only of potential economic value in a few areas. Each deposit will be discussed separately.

Manganese Nodules

To be of any economic value, manganese nodules must contain combined nickel and copper contents, the two metals of greatest value in them, of about 3 percent. The main area of the oceans where such deposits occur is in the so called Clarion-Clipperton Zone, between the Clarion and Clipperton Fracture Zones in the northeastern tropical Pacific. Here the nodules are abundant as well as of high grade. Very little of this area is, however, within any Exclusive Economic Zone. Nevertheless, study of the conditions under which high grade and abundant nodules form in the Clarion-Clipperton Zone has helped in locating other areas of the oceans where similar deposits might occur, and one of those areas which falls into an Exclusive Economic Zone is the subject of investigation under Imperial College's Marine Mineral Resources Programme.

Conditions leading to the formation of abundant high grade nodules in the Clarion-Clipperton Zone are essentially twofold. First there has to be a source of the metals, and second, an environment of deposition conducive to their uptake into the nodules. The source of the metals in the Clarion-Clipperton Zone is the abundant planktonic organisms that fall to the sea floor under the equatorial zone of high biological productivity. These extract metals from the surface waters and liberate them back into the bottom waters and interstitial waters of the sediments as the organisms decay and dissolve after death. These same organisms, by their decay, lead to a depositional environment within the sea floor sediments that promotes chemical reactions leading to the concentration of the elements in the forming nodules. Thus, by analogy with the Clarion-Clipperton Zone, other areas of abundant high grade nodules should occur under regions of elevated biological productivity where organic decay in the sediments promotes reactions leading to the concentration of metals in the deposits.

Most such areas occur in open ocean settings far removed from Exclusive Economic Zones, such as the central Indian Ocean and parts of the South Pacific. However, one of these areas impinges on the Exclusive Economic Zones of some western Pacific island nations, namely Kiribati and the Cook Islands (Fig. 6) and these areas have been the subject of study at Imperial College. An analysis of the environmental conditions within these areas led to the delineation of sites that should contain potentially economic nodules, and subsequent sampling of these sites confirmed their presence. Work in this area is continuing with detailed studies in small areas in order to determine the local variability of the deposits and the factors influencing this.

Cobalt rich manganese crusts

Ferromanganese oxide encrustations coat most exposed rock outcrops on the deep sea floor, but only in a few areas do they contain sufficient cobalt

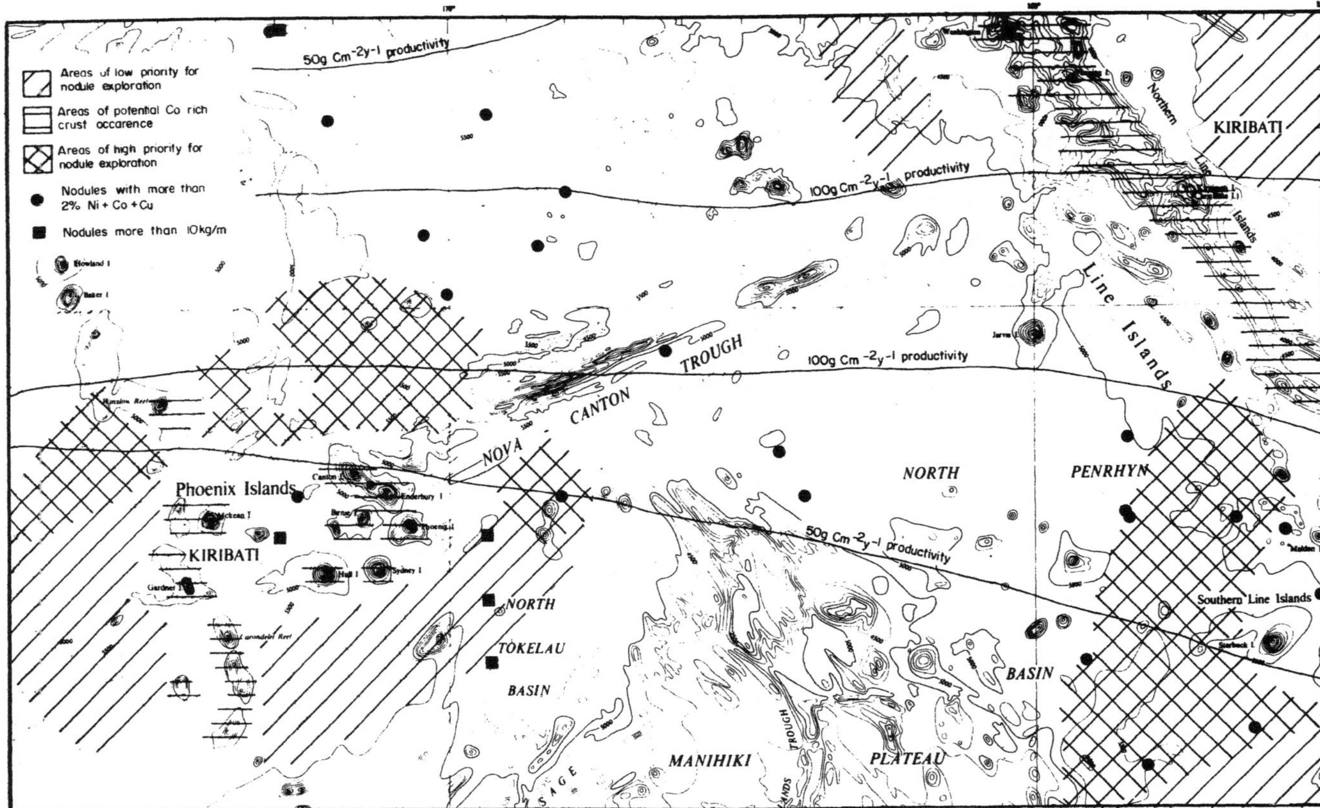

Fig. 6. Bathymetry, productivity isolines and areas of known or potential nodule and crust occurrence within Exclusive Economic Zones in the East Central Pacific Basin. (From D.S. Cronan, 1984, Criteria for the recognition of areas of potentially economic manganese nodules and encrustations in the CCOP/SOPAC region of the central and south-western tropical Pacific, South Pacific Marine Geology Notes, Vol.3, No.1, p.1-17; reproduced by permission of CCOP/SOPAC).

to be considered as a possible future resource. These areas are largely located in the equatorial and sub-equatorial western Pacific, and many of them fall within the Exclusive Economic Zones of the island nations of that region.

Cobalt enrichment in the crusts is concentrated in the approximate depth range of 1000-2000 m. Most seamounts do not rise to less than 1000 m from the sea surface and below 2000 m depth Co rich manganese crusts are rare. A whole host of other elements are also enriched in the shallow Co rich crusts relative to their concentrations in those from deeper water, but only nickel and possibly platinum are considered to be of possible future value.

Available evidence suggests that Co and associated element enrichments in the crusts is related to scavenging of metals from seawater by ferromanganese oxides within or adjacent to the dissolved oxygen minimum zone in the oceans. It follows that deposits are likely to be best developed where this zone is most intense. In the Pacific, this is in the equatorial and north tropical regions of the western Pacific, and nations with Exclusive Economic Zones in those areas (Fig. 6) such as the U.S.A. (by virtue of its trust territories) and Kiribati are undertaking active exploration programmes for the deposits.

Fisheries and other biological resources

by
J.A.Gulland
Marine Resources Assessment Group
Imperial College
London

Introduction

Before the recent major changes in the accepted extent of jurisdiction by coastal states over fisheries off their coastsa large proportion of the world's fish catch was taken beyond national jurisdiction. Even when catches were taken within national jurisdiction it often happened that significant catches from the same fish stock were also taken on the high seas. Thus it was impossible, in the case of most major fisheries, for a coastal state, or two or more coastal states acting together, to institute effective control of the exploitation of the resource. Where management was needed to prevent over-exploitation and damage to the resource, it had to be achieved through international cooperation, often in the form of a regional fisheries Commission.

The performance of these Commissions has been mixed. They did, and continue to do, a lot of good scientific work and did introduce a number of regulations, e.g. minimum mesh sizes in a number of North Atlantic trawl fisheries. These measures, however, usually fell short of what was needed for effective management of the fisheries. While it is true that when, for geographical reasons, a fish stock could be effectively managed by a coastal state on its own, management was not noticeably more successful, and some of the more spectacular collapses of fish stocks e.g. the Californian sardine, happened in stocks that were effectively under the control of a single state, dissatisfaction by coastal states with the performance of the international Commissions was an important factor in changing the Law of the Sea and bringing EEZ's into existence.

The introduction of EEZs has therefore altered fundamentally the pattern of the world's fisheries, to an extent that depends on the distribution of the fish stocks. It has significantly altered the balance between countries, especially between coastal states and countries operating long-distance fishing fleets. It has also altered, at least potentially, the way in which fisheries are managed, from what was essentially a matter of acting on the lowest common measure, acceptable to all participants in the fishery, to a situation in which the coastal state can apply whatever measure it thinks appropriate. At the same time there are a few resources, notably tuna and whales, which are available in significant quantities beyond 200 miles, and which have therefore been less affected by the introduction of EEZs, and present special problems. These points are discussed in the following sections.

The distribution of living resources

Fish, of a vast variety of species, are found in all parts of all the oceans, but it is only certain species in certain areas that occur in sufficient concentrations to support commercial fisheries. The differences in the basic production of plant life from place to place in the oceans are almost as great as the differences between forests and deserts on land. It is only the richer parts of the oceans that can support significant fisheries. Enough light to maintain the growth of plants only penetrates a few metres or tens of metres below the surface, and high productivity depends on having a good supply of nutrients into this surface layer.

The good fishing grounds of the world are thus largely restricted to two major areas, the continental shelves, where there is often, particularly in winter, complete surface to bottom mixing, and upwelling regions. Upwelling, bringing cool, nutrient rich water to the surface can occur in a number of situations, but much the most important upwelling areas are off the western margins of the continents - the Humbolt, California, Benguela , and Canary Current systems. In each of these areas the same pattern of consistent winds at an angle to the coast, which produce deserts on land produce riches in the sea. The shape of the Peruvian coast-line is particularly favourable to upwelling, and Peruvian waters have produced the world's biggest single species fishery.

Virtually all the shelf areas of the world lie within national EEZ's. The exceptions - the Flemish Cap, and the south-eastern tip of the Grand Bank off the Canadian east coast, and part of the Patagonian shelf around 40^[x^\S - provide only a very small part of the total world catch. The management of the stocks on these shelves present some fascinating biological and political problems, particularly in respect of how what is done in these areas affects what is done in the adjacent areas under national jurisdiction. They are, however,wholly untypical of present day problems.

Upwelling areas are not of necessity so closely linked to land masses, and there are upwelling areas in the central tropical Pacific, well outside any 200-mile EEZ, that are productive.These - the line grounds - were in the past important to sperm whalers, and are now of interest to tuna fishermen. The biggest upwelling systems depend on the winds moving the surface waters away from the coast, and these upwellings must be close to the coast. Though with the general off-shore movement of the water-masses the successive stages in the food chainare further and further from the coast, the main fishing grounds off Peru, California, or North-west Africa are still within 200 miles of the coast.

Thusnearly all the world's major fishing grounds lie close to land. In total it has been estimated that over 99% of the world catch is taken within 200 miles of the coast. This is not to say that now 99% is taken within national EEZ's. There are still places where for one reason or another - local political considerations, or arguments about the jurisdiction over the land territories - a 200 mile EEZ or equivalent form of coastal state jurisdiction over fisheries has not yet been declared. These include some productive areas - off Namibia and around the Falklands - which have been, and continue to be, important to long-range fleets. In these areas the pre-LOS conditions continue to apply.

More than the position where the fish are caught has to be looked at when considering the impact of EEZs on fisheries. All fish and other living resources move, even if some, like oysters, only move in the first few weeks of life. This movement, and the movement of water currents, mean that events in the EEZ of one country can be affected by what happens in adjacent EEZs or on the high seas. The extent of this effect on national fisheries, and on the ability of individual countries to manage their national fisheries without reference to what is being done by neighbouring countries, varies.

In many cases the effect is slight, being no more than the wandering of a few fish across the boundary. In other cases movements are substantial. These can be divided between seasonal migrations, and movements related to the growth and development of the individual. Seasonal movements are naturally most prevalent in temperate and sub-arctic waters where seasonal differences tend to be most marked, but do occur elsewhere. For example, many of the fish stocks off North-west Africa carry out substantial migrations, and because several of the countries have only narrow lengths of coast these migrations carry the fish in and out of the EEZs of a number of different countries.

Many fish species move during their life in a triangular pattern. From the spawning grounds the eggs and the larvae in the first few weeks of life drift fairly helplessly in the currents to arrive on the nursery grounds, where they may stay for some months or years before reaching a size at which they recruit to the main fishable stock. Then as they mature they may migrate to the spawning grounds to complete the cycle.

Depending on the details of the life-historythese migrations may take the fish into and out of the EEZs of several countries. For example salmon are spawned in fresh water but move to the ocean to grow. During their ocean life British or Canadian salmon move onto the high seas or into the coastal waters of Greenland. There they may be caught, raising arguments between those, including the Greenlanders, who want to catch them in the sea, and the Canadians or British, who point out the advantages of restricting harvesting to close to the spawning stream, where natural losses are least, and catches can be most easily adjusted to maintain the optimum size of spawning stock in each individual stream.

Similar arguments can arise in respect of most migratory fish which are not restricted to a single EEZ. Often biological considerations will point to one place as being the optimum point of harvest, with preferably little or no catches being taken elsewhere. Management of these species is a matter of finding an acceptible balance between these biological considerations and the natural wish of each country to take as much as possible in it's own waters with it's own fleet.

Because of these movements, therefore, there are many fisheries that cannot be managed by one country acting in isolation. This situation, however, differs in one important respect from that occurring in the pre-LOS era. The number of countries that are concerned is not open-ended. It is restricted tothose into whose waters the stock moves,and this is almost always small, and often no more than two or three. The problems of reaching agreement for effective action among this small number may be no more than that invoved in getting agreement between different groups in the same country.

Winners and losers

One of the expectations of the new law of the sea was that it would help redress some of the imbalance between rich and poor nations. One image that encouraged this expectation was that of large fleets of powerful distant-water fishing vessels from the developed countries devastating the resources off the coast of poor countries. This image was only partly correct. Certainly some distant-water fleets were causing damage to fish stocks which the coastal states could, under the current law of the sea, do little to prevent. The coastal states concerned were, however, not solely the poorer countries , nor has distant-water fishing been confined to the technologically advanced states (Tables 1 and 2).

TABLE 1

Catches (thousand tons) of fishing vessels fishing off the coasts of foreign countries, according to location of catch and nationality of vessel.

Nationality of vessel	Location: Off developed countries	Off developing countries	Total
Developed	10 716	4 572	15 288
Developing	430	641	1 071
Total	11 146	5 213	16 359

In 1972, the last year before extensions of jurisdiction by increasing numbers of coastal states began to move fisheries into the post-LOS situation of 200-mile EEZs or their equivalent, slightly more than a quarter of the total world catch was taken by non-local fishing fleets. Most of these fleets conformed to the popular image, coming from the traditional fishing countries of western Europe (which have been fishing off north-eastern America since the time of Columbus), Japan, and the USSR and other countries of eastern Europe. In addition, however, several developing countries including the Republic of Korea, Cuba, and Ghana operated fleets of large distant water fleets, while smaller trawlers from Thailand were operating throughout much of south-east Asia from India to Vietnam to Borneo.

The locations of the distant-water catches depart much more markedly from the popular image. As Table 2 shows, nearly all the non-local catches came from six of the fifteen statistical areas in which FAO records the world catches. These are the rich and extensive areas of continental shelf of the north Atlantic and north Pacific, and the upwelling areas off north-west and south west Africa. Among the biggest winners from the changes inthe law of the sea were therefore Canada and the United States. Several western European countries also gained, although around the North Sea the existence of the EEC, and the authority given to the

European Community under theTreaty of Rome and the development

TABLE 2

Catches in 1972 by non-local vessels in different areas

Area	Total catches '000 tons	Non-local catches '000 tons	% of total
N.W.Atlantic	4 327	2 292	53.0
N.E.Atlantic	10 699	3 667	34.4
E.C.Atlantic	1 488	143	9.6
Mediterranean	1 165	40	3.4
S.W.Atlantic	805	24	3.0
S.E.Atlantic	3 013	1 771	58.8
W.Indian Ocean	1 809	201	11.1
E.Indian Ocean	821	88	10.7
N.W.Pacific	14 531	2 936	20.2
N.E.Pacific	2 774	2 254	81.2
W.C.Pacific	4 770	479	10.0
E.C.Pacific	982	287	29.2
S.W.Pacific	275	199	72.3
S.E.Pacific	5 539	48	0.9
TOTAL	56 109	16 539	29.4

of the Common Fishery Policy has complicated the issue, and reduced the extent to which it is possible to say that one particular country has gained or lost. Off west Africa the political confusion in Namibia and ex-Spanish Sahara, off which some of the richest grounds lie, has meant that the inhabitants of these regions have received few of the benefits that they might otherwise have enjoyed.

The British experience has been mixed. The English distant water fleet, which was largely responsible for opening up the North Atlantic trawling grounds in the first third of this century, (the traditional cod fisheries from the fifteenth century onwards was by hook and line), and which supplied the bulk of British fish supplies in the 1960's, has lost access to it's main grounds, and has almost vanished. Since the coastal states concerned - Norway, Iceland, USSR, Greenland and Canada - were fully capable of taking up any slack left by the departure of the distant water fleets, there was never much chance of this fishery continuing for more than a brief phasing out period. On the other hand the U.K. did gain control over rich resources in the North Sea and to the westward. Though this control is modified by the Treaty of Rome and the powers of the EEC, some fishermen, particularly Scots are clearly benefitting e.g. in the supply of mackerel to Russian vessels which previously would have caught the fish themselves.

Events since 1972 in the areas of significant distant water fishing have been varied, depending on the conditions ashore. In the south-east Atlantic little has changed. Most of the distant-water catches are taken off Namibia. Here no 200-mile limit has been introduced, and none is likely to be introduced so long as few countries recognize the South African regime, and no more acceptable regime is installed. Management still lies with an international Commission of the traditional pattern. This Commission - the International Commission for the Southeast Atlantic Fisheries has performed as well as other similar bodies, and is perhaps most notable as probably the only smoothly operating international body in which two of the most active members are South Africa and the USSR.

Even when the coastal state has extended jurisdiction, this has not necessarily meant the end of substantial distant-water fishing. The coastal state may well find it convenient to allow foreign fishing to continue under suitable conditions. Thus in the Bering Sea and Gulf of Alaska fishing by U.S. fishermen has been largely confined to the most valuable species - salmon, halibut, crabs and shrimps - leaving the hard, uncomfortable, often dangerous, and less economically rewarding work of catching and processing the lower valued species - Alaska pollock, flounder etc - to Asian fishermen, though some American fishermen have made very good money catching pollock for direct delivery into Russian factory ships. For different reasons the coastal states of north-west Africa have been slow to develop enough catching capacity to replace more than a small fraction of the foreign fleets that were fishing there in 1972.

Only in the north Atlantic has the balance in catches between local and foreign fleets altered dramatically in the way many people expected. In eastern Canada, Iceland, Norway and elsewhere there were already well-developed local fishing industries, which could easily take up the slack left by the departure of the foreign fleets, especially where, as was the case for many stocks, there was a need for a reduction in the total amount of fishing. The extension of limits was often followed by a temporary drop in total catch, but this generally arose from a need to rebuild depleted stocks, rather than any failure to expoit stocks to their full potential.

The future of fishery management

Although the changes in world fisheries that have so far occurred may appear to have had an impact on world fisheries that is much smaller than was expected, the long term impacts may yet turn out to be fundamental, not only in respect of the balance between coastal states and distant-water fisheries. To see why this should be it is necessary to look in more detail at recent principles and practices of applying management measures.

Because any measures introduced into international fisheries have to be acceptible to all participants, they have to be such that no participants lose, and if possible all gain, if not necessarily equally. Also, to persuade everyone that this is indeedthe case, the scientific evidence has had to be clear and unequivocal. The need for consensus has meant that the grounds for management action have tended to be limited to biological criteria, especially the conservation of the stock, to the exclusion of economic or social criteria. Where the latter have been considered, it has generally been to resist the introduction of new measures rather than to propose action. This has been true even in those fisheries where the potential economic benefits frome.g. reduction of over-capacity, greatly exceed those from improvements in total catch. The same principles have tended to be applied to fisheries under single national jurisdiction.

These constraints have severely limited the number of situations in which management measures have been successfully introduced. These are principally those of a single fairly homogeneous fishery, or a small number of similar fisheries acting on a single fish stock. Even then the measures have tended to be limited to those that do not interfere too much with the day-to-day operations of the fishermen, e.g. mesh regulations.

Typically, it was only after the collapse of the stock concerned e.g. the North Sea herring, that it was possible to agree on more drastic measures such as limits on total catch or total effort. An early exception to this was the International Whaling Commission, which during the 1950's limited the international catch of Antarctic whales to substantially less than the fleets of those days could have taken. Unfortunately the supply of scientific advice at that time was not good enough to show that those limits were still too high. By the time better advice was available in the mid-1960's the reduction in total catch was too big to be easily acceptable. The subsequent collapse of the whale stocks is well known.

Although controls on the total amount of fishing, through the imposition of Total Allowable Catches(TACs) for the major species, and the allocation of shares in these TACs to individual countries is part of the current management system for the North Sea under the Common Fishery Policy, in practice the level of TACs for which agreement has been possible represents no more than a limit on further expansion. These limits prevent - or should prevent - the situation getting worse, but is little advance on the pre-EEZ situation.

Present practices in many areas, including the waters around the U.K. fail to address adequately two major problems facing fisheries today - the great over-capacity in many fisheries, and the interactions between fisheries on different stocks. Simple economic theory predicts, through what has been called the tragedy of the commons, that all open-access fisheries will suffer from over-capacity, and experience has shown theory to be in this case almost completely correct. Though the introduction of EEZs has limited access for most resources to one or a handful of countries, within each country fishing has been, at least until recently, an occupation open to all. Since the degree of over-capacity increase with higher prices and greater technical efficiency, the free market mechanisms, by themselves fail to cure the problem, and if anything make it worse.

There are a variety of ways of dealing with over-capacity, ranging from the traditions of the Pacific islands, and elsewhere, which set out who can catch fish when and with what methods, to sophisticated methods of licencing and limitations on entry. Though all involve considerable changes to current fishing practice, and can seldom be introduced without long arguments and discussions with all parts of the fishing industry, limited entry schemes, of one type or another, aare now in place in several parts of the world, from Western Australia to Alaska. Around the North Sea, however, despite the leading part played by scientists at Lowestoft and other institutions around the North Sea in developing the biological aspects of the theory and practice of fishery management, little has been done, on either the theoretical or practical side, about moving towards a significant reduction in the excess capacity that exists in the area, and which present the opportunity for very considerable economic benefits, particularly in the long term.

The problems of the interactions between fisheries are only slowly being tackled. It has long been recognized in principle that fisheries and fish stocks do not exist in isolation, and that each interacts with other fisheries and stocks in the same area to a greater or lesser extent. These interactions can be operational i.e. the capture of one species by fisheries targetting on other species, or biological e.g. through the predation of one species on another or competition between species for the same resource.

The technological interactions have been the easier to observe and to deal with. Limitations can be put onn the amount of non-target species that can be landed and sold, and quotas in fisheries targetting on one species can be adjusted to take account of the expected catches of that species in fisheries targetting on other species. Special measures can be taken to protect specially vulnerable or desirable species. For example, theCalifornian fishermen catching tuna with purse-seines in the eastern Pacific have had to make considerable changes to their gear and the methods of operating it in order to reduce the incidental killing of porpoises in their nets. Less attractive or less important species may be less fortunate. In the Irish Sea the common skate is vulnerable to most fishing gears, and having a very low reproductive capacity, can withstand only a very light fishing pressure. The indications are that the Irish Sea population of skate is declining towards extinction.

The technological interactions,with some exceptions such as the skate, are in most cases relatively minor. The biological interactions though less immediately obvious, may in the long run be more important. For example, one of the more bitter public arguments over the management of marine resources in the U.K. arose over the possible desirability to control grey seals in the interests of fishermen with whom they were competing for salmon and other fish. A more complex situation of essentially the same kind arises from the competetion between fishermen and cod for small species such as Norway pout and herring. Because cod eat herring it is probable that catches of herring and other small fish would be increased by reducing the abundance of cod by deliberate overfishing. Interactions when one or both species are very young could have particularly dramatic effects. Since 1960 the recruitment of young cod, haddock and other bottom-living species has greatly increased, and the catches of these species has increased from a remarkably steady level of a little under half a million tons to 1 to 1.5 million tons. This increase took place just as the stocks of herring and mackerel had collapsed from over-fishing, and there are signs that as the herring stocks are recovering the recruitment of cod is declining. This suggests that it would be desirable, from the point of view of those fishing cod or haddock, that stocks of herring and mackerel should be kept at a moderately low level.

A common feature ofall these interactionsis that their magnitudes are highly uncertain. There was little disagreement that seals ate fish, some of which were of commercial species. What was at issue was whether the amount of commercial species, especially salmon, was high enough to have a significant effect on the fisheries. Another common feature is that there would be losers if the management actions implied by the analyses of the interactions are carried out. Either fishermen or those wanting to protect seals have to sacrifice some of their interests in favour of the other group. Increasing herring catches by reducing cod predation would require cod fishermen to catch less cod at greater cost, and so on.

These are the sort of measures that would be difficult or impossible to consider under the consensus condititions existing before EEZs. Balancing interests and accepting uncertainties are typical of many problems that governments have to deal with. Under the new regime it seems reasonable that governments with direct jurisdiction will be able to act more effectively on these sort of problems. They should also be able to act more effectively on matters concerning economic performance. Overall, this greatly widens the range of management problems that can be dealt with.

The significance of this widened range for fisheries will depend on the relative benefits that could have been obtained under the old regime, and management by consensus compared with those that are only likely to be obtainable under different decision making processes. A quantitative judgement would be difficult to make. A subjective judgement is that, at least for the fisheries around the British Isles, the total benefits that might be achieved by management schemes that took full account of all likely interactions, (with due allowance of uncertainties), and of the long term economic consequences of different actions, could provide economic and social benefits that greatly exceed any that have been achieved by management actions taken so far, or which could ever have been achieved under the old regime.

Special cases

In the text of the law of the sea convention two groups of species are given special mention - marine mammals and highly migratory species. The reasons for this special attention and the implications for future fishery management are worth examining in a ltle detail.

A cynic might define highly migratory speies as those caught by Californian fishermen. Certainly the development of the concept owed a lot to the need of the U.S. to harmonize the interests of New England fishermen, who felt threatened by large fleets of distant water fleets off their coasts, and of Californian tuna fishermen, who took much of their catches off the coast of Mexico and other Central American states. The concept did, however, also recognize biological realities and the difference between purely local stocks, and those whose movements carried them into and out of the waters of several states. The division in the law of the sea text is, however, more on the basis of taxonomy and political convenience, rather than the actual extent of migration. Thus all tunas, some of which do not move over long distances, are included, while the herring that migrate from Norway to Iceland, and some other long-distance migrants are excluded.

The thrust of the sections of the L.O.S. convention dealing

with highly migratory species is to call for close cooperation between all states concerned. In effect this is little more than a continuation and strengthening of the international arrangements typical of most large fisheries in the pre-EEZ era. Where much of the catches from the stocks concerned are taken or can be taken beyond 200 miles the changes in legal regime have indeed had little impact. The two major international bodies specifically concerned with tuna - the Inter-American Tropical Tuna Commission, and the International Commission for the Conservation of Atlantic Tuna - are continuing their work much as before. Their strengths - compilation of reliable and comprehensive statistics, and good scientific analyses - and weaknesses - inability to introduce strong and effective management measures - have continued. Efforts have been made to revise I-ATTC, or to replace it with another body which pays more attention to the interests of coastal states, but without much success.

In the south-west Pacific the picture is very different. Here the mosaic of islands is sufficiently closely spaced that the 200 mile zones cover most of the region, with only a few small and isolated areas beyond 200 miles. The result is that any vessel exploiting the rich tuna resources of the South Pacific must spend most or all of it's time within the 200-mile limit of one or other country.

The catches of tuna in the region, though less than the bulk catches by long-range vessels in the North Atlantic or some of the upwelling areas, are quite considerable and with the high price of tuna the value is very high in comparison with the total GNP of most of the countries concerned. With the exception of Papua New Guinea, and to a lesser extent the Solomon Islands, these are all small islands and their land area is a minute fraction of the area of their EEZs. If they could take full advantage from the tuna fishing in their EEZs this would have a very significant effect on their national economies.

This has not been easy. The tuna vessels come mainly from Japan, with others from Korea, and more recently the U.S. They are largely owned or controlled by a handful of large companies. These can command skilled negotiators and the early post-LOS negotiations, which were between individual distant-water fishing countries and individual coastal states resulted to agreements which were definately favourable to the former. The island states then recognized that they had to work very closely together, and the cooperative arrangements developed steadily. The structure for such collaboration already existed in the South Pacific Commission, to which all territories in the region, including the metropolitan states, belong, and the South Pacific Forum which is restricted to the independent states in the region.

The SPC has provided a regional compilation of statistics and carried out research on the distribution and abundance of the tuna. This work has been very valuable, though it involves no new approach to regional cooperation, being very similar to the work done by bodies in the North Atlantic and North Pacific. A more radical approach has been taken by the Forum. This has established a special Forum Fishery Agency, which has much greater powers regarding policy and operational matters than any similar body. In particular the work of the FFA has enabled the island states to present a uniform and integrated approach in negotiation with the tuna fishing countries, which hasbeen reflected in much more favourable terms e.g. on the level of licence fees in the agreements to allow access to the region. The FFA has also taken actions, e.g. in the preparation of a region-wide register of vessels granted access, which go some way beyond what has been the traditional role of a regional body. Though the regional arrangements for tuna in the South Pacific are in some ways exceptional, they do provide one case where the outcome of the changes in jurisdiction are going a little way towards redressing the balance between rich and poor that was hoped for by some of those working for changes in the traditional law of the sea.

The special attention paid to marine mammals is an indication of another way in which fisheries and other marine affairs can no longer be carried without regard to external opinion. Marine mammals have become symbols of the conservation movement, and tests of man's ability to use the environment wisely. They receive attention that is far beyond their importance in narrow economic terms. Members of the European Parliament received more letters from their constituents about the proposal to ban the import of certain seal products into the EEC than about any other subject. Some of this public concern can be attacked as being emotional or illogical. For example virtually all the attention over seals was focussed on the Canadian harp seals, which numbered over a million and were probably increasing even before the drop in catches in the last couple of years, and virtually none over the monk seal, which occurs in European waters, and is in clear danger of extinction.

There are some good reasons for special public attention. Whales are the biggest animals that ever lived, and dolphins may well be among the most intelligent. Marine mammals are also, by virtue of their long lives and very low reproductive rates, particularly vulnerable to over-exploitation. Poor management is, therefore, likely to have particularly damaging effects. The present calamitous state of many - though not all - whale stocks in the Antarctic and elsewhere is due more to the biology of whales rather than to the wickedness of the whaling industry or the inefficiency of the International Whaling Commission. Both of these behaved, if anything better than the corresponding groups in fisheries in other parts of the world.

Marine mammals do therefore require rather a careful and conservative approach in their management. A sufficiently careful and conservative approach is, to be realistic, somewhat unlikely within the traditional framework of fishery management, with strong weight given to immediate economic interests. Arrangements that give some attention to other interests e.g. of conservation groups should therefore be welcomed. At the same time the principles of special treatment for marine mammals, most clearly expressed in the U.S. Marine Mammal Protection Act - the LOS text is not specific on how they should be treated - and for greater involvement of wider interests do themselves raise new and possibly difficult issues. Are marine mammals receiving too favourable treatment, and do outside interests have too much influence?

It is easy to find possible examples. The IWC has, in responseto calls for a moratorium introduced zero quotas for allcommercial whaling. These are to be reviewed not later than 1990, but it is hard to see, with the IWC's current membership and voting patterns, any proposal for altering these quotas for any stock receiving the necessary three-quarters majority even if the scientific evidence showed clearly that the stock concerned could sustain the catch rates proposed. At such a time the Commissions structure will come under serious strain.

Several stocksof marine mammals e.g. grey seals, are recovering from many years of heavy exploitation. Many of these stocks interact with fisheries, though the exact magnitude of this interaction is very difficult to measure. While the mammal stocks were still small the fishermen concerned could perhaps accept anylosses as just one more element in their struggle to make a living. As numbers increase the call by fishermen for controls could well also increase to a point at which the government concerned could be squeezed between fishermen and some elements of the environmental movement. For example American fishermen along both the Atlantic and Pacific coasts are increasingly complaining about seal damage, and looking for some relaxation of the MMPA.

In themselves these problems are relatively minor. The whaling industry is now, and under any circumstances will remain, a shadow of it's former size. Only a small proportion of American fishermen suffer sufficient visible damage from seals to make strong protests. They can be seen, however, as early indications of a trend that will surely continue. Just as the establishment of EEZs has extended the geographical scope of the powers and responsibility of coastal states, so the widened public awareness of the oceans, and what is happening in them will widen the range of interests and pressure groups that the coastal states will have to take into account in deciding how to manage the resources of the EEZs.

OCEAN RENEWABLE ENERGY RESOURCES : A CHANCE FOR THE FUTURE ?

Philippe MARCHAND, IFREMER
66, Av. d'Iéna 75116 PARIS

The ocean offers a considerable potential of renewable energy. The sources are numerous : tidal energy, wave energy, ocean thermal energy, salinity gradients, current energy and biomass. Wind energy is not typical of oceans. One can observe than despite this large potential, the only industrial exploitation of marine energy is the famous Rance tidal plant which was built twenty years ago, before the oil crisis. During the last ten years, a big work was done in order to establish the technical and the economial feasibility of marine energy and to promote pilot plants, mainly in OTEC and wave energy.

Unfortunately for this research, the recent slump of oil prices will freeze initiatives and research budgets. And probably for a long period. Economic priority is not presently in the energy area but in other new fields like communications which is a sophisticated form of energy.

This presentation will be focused mainly on waves, OTEC and tidal which are the most promising forms of marine energy. It will present :

- resources,
- state of art,
- cost and market,
- findings.

I - THE RESOURCES

The ocean covers more than 70 % of the planet and represents a very large reservoir of energy. This energy is coming from two phenomena :

1. gravitational forces which give resonance effect in certain basins (generating tides),

2. solar radiations : $5{,}5 \cdot 10^{16}$ W is the power given by the sun to the oceans, half is converted into evaporation which generates winds, creating waves and majority of currents. Wave energy which is finally dissipated along the coast is low : about $2 \cdot 10^{12}$ W which represents 1/5 of the power corresponding to the world total energy demand. So mechanical dissipation of energy (waves, currents) is very low regarding oceanic thermal transfers which regulate repartition of solar energy and maintain the equilibrium of climates.

OTEC uses the difference of temperature between surface tropical waters and deep waters, the theoretical production would be ten times greater than wave energy production.

The following table gives order of magnitude for the various marine energy forms.

RESOURCE AND DENSITY OF MARINE ENERGY

	Theoretical Production (GW)	Density (in m of water)
TIDAL ENERGY	100	10
CURRENTS	300	0.05
WAVES	1,000	2
SALINITY GRADIENTS	3,000	240
OTEC	10,000	570

The world energy demand corresponds to 10,000 GW (or 10,000 large nuclear plants).

We can observe that OTEC presents both the most concentrated form of energy and the biggest theoretical potential. More specifically, we can characterize the different energy sources :

Tidal energy : the resource is limited to about twenty good sites with tidal amplitude exceeding 5 m. The production can be predicted with a very good precision because of the astral origin of the tides. The Severn estuary is probably one of the best site capable of 12 GWe.

Wave resource : very variable energy (0 in calm waters, 1,000 KW/m during a big storm), no prediction is possible and the production has to be injected into the grid when it is available. The total wave resource of EEC (90 GW) would correspond to 70 % of its electricity consumption.

Otec : the resource is very broad : one third of the ocean is concerned by a difference of temperature exceeding 18°C between warm surface layers (acting like a huge solar collector) and cold deep sea water (4° at 1,000 m depth) originated from polar regions. The resource is stable, with little seasonal fluctuations.

Currents : various types of currents among them tidal and permanent ones like the Gulf Stream. Energy density is very low according to Bernouilli law : the most rapid currents correspond to a very low head, under half a meter.

Salinity gradients : 1m^3/s of fresh water mixed with sea water produces 2.2 MW of chemical energy. All estuaries are theoretically capable of an energy production. For example, one tenth of the Congo flux would produce 3 GW.

II - STATE OF THE ART

the status of the technology is quite different between ocean energy sources, from the most commercial until the most speculative, one can make three classes :

1. Industrial level : tidal energy is a reality with the 240 MWe Rance unit in France. No major developments would be required to build, for example, the Severn barrage or a big plant in the Bay of Fundy. Turbines are disponible, technology of caisson is proved.

2. Demonstration level : Otec technology has to be demonstrated with a large pilot plant. Critical components are :

- the cold water pipe : how to build, to deploy such a large diameter pipe ? Successfull partial tests were done until 2,5 m in diameter. Two materials are emerging from the research : the polyethylen for small diameters and the FRP sandwich for larger pipes. For floating plants, one of the technical challenge is the connection of the suspended pipe. For land based plants, the main challenge is to lay a long pipe (typically : 3 kms) above the irregular volcanic slope of the tropical islands. The work was completed for small diameters (70 cm) only, in two Japanese experiments (Nauru and Tukunoshima).

- heat exchangers : for closed cycle, the technology is derivated from classical units, an adaptation for OTEC has to be done, taking into account biofouling control, enhancement of heat transfert and scaling-up. For open cycle, the problem is the size of turbine and how to maintain the vacuum in the exchanger.

- floating platform : the conception of the platform is classical but mooring disposals and cable to transmit electricity need to be demonstrated in more than 1,000 m water depth.

Presently, a few little OTEC plants were built to demonstrate the technical feasibility, but the economical viability has to be proven through a large demonstration plant. Three countries have realized pilot plants : Japan with the 100 KW Nauru plant (closed cycle, land based) which was in operation during nine months en 1982, USA with Mini OTEC (50 KW) and OTEC 1 (1 MWe without turbine) in 1980 in Hawaï, France with a 12 KW plant in Cuba half of century ago.

Wave Energy : numerous concepts for wave energy conversion systems were developped during the last ten years mainly in Great Britain. Only two major at sea experiments were done : one concerns the "KAMEI", a wave energy vessel using the principle of the "oscillating water column" which was the subject of an I.E.A. collaborative experiment in Japan and connected to the mainland grid, and a demonstration on the same principle but onshore, in Norway. The main technical problem is to build a structure storm resistant at an acceptable cost. A lot of ideas are able to work but the question is : at what cost ?

3. Feasibility level

Other ocean energy sources (currents, biomass, salinity gradients) were experimented in laboratories only and the at-sea feasibility has to be proved. For salinity gradient, the main technical problem is relative to the osmotic membran, an other problem is the location of the disposal in estuaries where the maritime traffic is generally developped.

Concerning biomass, the main technical problem is the exploitation of non indigen species of algae like Macrosystis pyrifera. Concerning the captation of marine current, the main question is to conceive a very low head marine turbine.

III - COST AND MARKET

The cost of marine energy is closely site dependent and generally speculative in absence of major industrial realization, excepting the Rance tidal unit. Nevertheless, it is possible to give some elements for OTEC, waves and tidal energies :

Tidal energy : For a good site like the Rance, with maximum tide range of 11,5 m, 240 MWe and an annual production of about 500 GWh, the total production cost of KWh was 0,17 FF in 1982. If the plant was built today, with the technics used 20 years ago, the cost of KWh will be of about 0,40 FF, comparable to the cost of coal electricity.

For other good sites like the Severn estuarie or the bay of Fundy, cost estimations give comparable figures. The main problem for such big installation is political because the capital cost is high, the construction time is long, the regional impact is great. For exemple : the Severn "inner barrage" (7 200 MW and 13 TWh/year) will cost 6 000 M£ (1980 prices) and construction time will be of 12 years. What is the government which will decide such an investment, the result of which being visible more than 10 years later ?

Wave energy : Realistic economic assessment of wave energy utilization systems is difficult as no units are commercially available. Devices included in the U.K. assessment for a 2 GW scheme where costed between 8 and 23 pences/KWh so at least 1.6 time above the attractive cost for wave energy utilization (5 pences/KWh). It is the reason while research in W.E.C. is not presently very active. But for medium scale applications, concepts like the oscillating water column or the overtopping reservoir could be competitive for remote location where cost of electricity is often 2 or 3 times higher than in major developping countries.

It is very difficult to speak of an export market without a conclusive demonstration plant. An I.E.A. Report (Berge, 1983) estimates that a total world market for wave energy devices could be worth $ 7 billions ant this could be increased after successful demonstration.

OTEC : The cost of OTEC is particularly sensitive to real interest rates (r) as it is a capital intensive form of energy. Annual operating expenditures account for no more than 10-20 % of total depreciation, which therefore weighs heavily. Estimated costs are given in the following table :

Net Capacity MWe	INVESTMENT $/KWe (1980)		Average cost of electricity (cent/KWh) r = 5 %	r = 10 %
5	12 000	20 000	17	25
10	8 200	14 500	11	17
40	4 500	7 700	6	9
100	3 200	5 900	4	7

The cost of electricity is also highly dependent of the quality of the hot and cold source : cost increases by 50 % when the temperature difference falls from 22 to 18°C.

OTEC has good chance to be competitive when compared with diesel-based electricity which is at present the main source of energy in the Island market and for small and medium-sized plants, if the oil cost is exceeding 30 $/bbl. The recent slump of oil prices will delay the hope of competitiveness, and the need of a demonstration in the short term. Unfortunately, it will probably frozen for several years initiatives and research budgets.

Concerning the market, a recent french study estimates the market to a potential of 3 500 MWe (of the 30 000 MWe to be installed) in the 20 countries with the most favourable OTEC sites, i.e., 120 OTEC stations with a unit capacity of less than 50 MWe. Much more optimistic figures were given by US study, including large floating OTEC plants, for the 70 tropical countries and territories concerned.

—

IV - FINDINGS : OCEAN ENERGY A CHANCE FOR THE FUTURE ?

The resource is very large, the technology is ready for demonstrations but the competitiveness remains problemistic. It is a major obstacle, with the recent slump of oil prices and high interest rates.

The three main ocean energy sources are site complementary with OTEC for tropical and equatorial areas, waves for medium latitudes (and also tropical alizean sites) tidal energy sited in 20 good locations generally with a large energy demand

If we except tidal energy which was commercially demonstrated with the unique Rance unit, OTEC and wave energy conversion (WEC) technologies are still in infancy and no devices are commercially available. The United States, Japan and France presently lead the world in the development of OTEC technology, the United Kingdom and Norway are leaders for wave energy conversion.

Intensive R and D efforts are required before a practical commercial OTEC or WEC unit could be built. Demonstrations, even with present technology, are technically feasible and will be very profitable if it could occur in the next years. Demonstrations on large scale are therefore costly and there is a reluctance by governmental and commercial interests to invest large amounts of capital in such high risk ventures.

The present economical environment is not very favourable to go ahead with the large slump of oil costs and high interest rates which handicap capital intensive forms of energy. If ocean energy sources like tidal energy, OTEC and WEC remain of great interest for the long term, the present economic environment dissuade public R and D allocations.

Nevertheless, it is highly desirable to maintain R and D efforts even if the demonstration goal is delayed and less compatible with the classical duration of governments : long term is classically not urgent for politicians because they will not profit of the result during their public life. It is the reason while supranational bodies like EEC, or international financial institution able to persit in a long term research effort, have to get interest in ocean renewable energy sources.

—

BIBLIOGRAPHY

- Commission des communautés européennes, Energie Thermique des Mers. Evaluation pour la CEE, Projet de démonstration, rapport EUR 9895 FR, by Philippe MARCHAND, CEE, Brussels, Luxembourg, 1985, 299 p.

- Commission of the European Communities, Wave Energy ; Evaluation for EEC, reference : XVII/BR/83/495, by Tony Lewis, CEE, Brussels, May 1984, 137 p.

- Vingt ans après la Rance, une expérience marémotrice, by P. Hillairet, in La Houille Blanche, n° 8, 1984, 10 p.

- Les énergies des mers, in Oceanis, Vol. 8, Fascicule 3, pages 161-293, 1982.

- Les énergies des mers, in Océanis, Vol. 11, fascicule 7, pages 629-718, 1985.

MODERN SURVEYING TECHNIQUES

Introduction

The title of the paper I and Dr Laughton have been asked to present was selected long before either of us started to put our contributions together, or even to decide in detail what to talk about. When we did begin collecting our ideas together, we decided, rather than to deal with surveying techniques as a subject in their own right, to look a little deeper, at the demarcation of EEZs, the rules and criteria for deciding their limits, and how surveys can be carried out to determine first which of the many different criteria are more advantageous to the country seeking to declare an EEZ, and then what limits it can actually claim.

The subject falls naturally into two parts:-

"Geomensuration" - heights of tide, distance offshore, depths, and so on, which I will be dealing with;

and "Geophysics" - angles of slope, thickness of sediments, etc, all closely allied to the detection of the possible existence of mineral resources, which Dr Laughton's paper will be dealing with.

Before getting down to the meat of my paper I should here enter a disclaimer. I am in no position to express any opinion on the provisions of the United Nations Convention on the Law of the Sea, on the United Kingdom's attitude to the Convention, or on the relation between it and the body of customary law on delimitation now growing up. I am simply going to try to discuss the surveying to establish the physical facts which needs to be done before the lawyers can start preparing their briefs.

Delimitation

Let us start at the beginning, with the baselines from which the breadth of the Territorial Sea, the Contiguous Zone and the EEZ are all measured.

Article 5 of the Convention states, clearly enough, that "Except where otherwise provided ....., the normal baseline ..... is the low-water line along the coast as marked on large-scale charts officially recognized by the coastal state." So far, so good, though as I shall have to mention later when dealing with tides, just what can be sensibly shown as "the low-water line" can itself need extensive observations in some cases.

Article 6 moves on to consider islands and atolls with fringing reefs, where the baseline is defined as "the seaward low-water line of the reef, as shown by the appropriate symbol on charts officially recogni̇zed by the coastal state". Here we already have the seeds of confusion. Where there is a well-defined reef edge this is what is normally shown on the chart, but it is by no means always the accurate low-water line, which can lie some distance to land-ward of the reef edge. In practice, I believe that the reef edge is normally taken as the baseline. This may be academic for a 3-mile or 12-mile Territorial Sea, but when we are looking at EEZs, and the reef-edge is less than 400 miles from another atoll belonging to another sovereign state the academic point can suddenly become of practical significance.

Sticking for the moment to features controlled by the level of low water, Article 13 treats of "Low-tide Elevations" - features "surrounded by and above water at low tide but submerged at high tide". Here the problem is to make an accurate determination of whether a bank or rock is actually "above water at low tide" - in other words, above Chart Datum - or not. Some of you may remember the case of the pirate radio station on one of the Thames forts where the juridical competence of the UK courts to deal with the case hung on the precise status of a bank in the estuary, and where the clinching argument was provided by a photo of one of our surveying officers in seaboots standing on sand which, though distinctly wet, was equally distinctly just above the level of the very calm sea round him!

Here, too, we come to the first time that distance matters. "Where a low-tide elevation is situated wholly or partly at a distnce not exceeding the breadth of the territorial sea from the mainland or an island, the low-water line on that elevation m ay be used as the baseline .....". So the surveyor needs to be able to establish accurately not only the low-water line of both mainland and offshore banks, but the precise distance between their nearest points. Small changes in distance or elevation can obviously have far-reaching effects on the area of Territorial Sea claimed, and even greater ones on EEZ area. Nowhere in the Convention are any criteria for determining the low-water line stated, only the simple statement "as marked on large-scale charts officially recognized by the Coastal State". This means, effectively, that we are talking about elevations above Chart Datum. But Chart Datum is itself merely an arbitrary level "so low that the tide will not frequently fall below it" - to quote an International Hydrographic Organization resolution on the subject. Modern Admiralty charts and surveys use Lowest Astronomical Tide - the lowest predictable level of tide under average meteorological conditions -, but even to determine this accurately a considerable period of tidal observations are needed on which to base the predictions from which the level is derived.

Perhaps fortunately for the surveyor - at least it makes our job more interesting - God has not got a tidy mind, and He did not make all the coastline of the world in a series of neat, regular areas trending in any particular direction. Coasts may be heavily indented and broken, and the Convention deals with this possibility in three ways: by the drawing of straight baselines along the outer edges of a range of features, by utilizing a regime of closing lines across the mouths of bays and rivers, and by the application of special rules for Archipelagic States.

The first need not concern us unduly - Article 7, the article in question, is practically an open invitation to the drawing, by States, of fairly arbitrary baselines governed more by the general "look of the thing" than by any strict determination of lengths and distances.

When we come to Bays, however, we start to need precise measurement of area and length. A "bay" is defined as an indentation whose water area is larger than the semicircle whose diameter is the line drawn across the mouth of the bay. So we need a measurement of the area enclosed by any bay-closing line before we can see whether we are entitled to draw such a closing line. Further, since such closing lines are only allowed to be up to 24 miles long, we need to know, in the case of an extensive bay, at what point its sides close to the magic 24-mile length.

The concept of an "archipegaic state" is a new one in the Convention. Though in considerable degree the definition is a common sense one, nevertheless there are again quantitative elements in the criteria which require accurate surveys to prove. The streight archipegagic baselines from which the state may start to extend the limits of its Territorial Sea and EEZ obviously enclose a mixture of land and water. The ratio between land and water must not exceed 1 to 9 - nine times as much water as land, but the whole area of a particular atoll within its fringing reef may be counted in the "land" side of the proportional equation.

As well as the area criteria, there is also a length criterion. Baselines are not to exceed 100 nautical miles, except that up to 3 per cent of the total number of baselines may stretch to 125 miles. Strangely enough there is no similar length criterion in paragraph 7 of the article defining an atoll, though there is a geological one in that limestone islands are specified rather than islands in general.

All these problems, and the surveying solutions of them, are comparatively easy when dealing with island states more than 400 miles from their nearest neighbours. Unfortunately there aren't many of them. When states are adjacent on the same land mass, or their EEZs meet across a stretch of water less than 400 miles across, the precise positions of the claimed baselines, and hence the ability of the data on which the claims are based to stand up to rigorous scrutiny, becomes a matter of great importance, on which the possession of considerable mineral and living resources can stand or fall. So modern surveys which not only are accurate but which can be seen to be accurate are essential.

Surveying

The sine-qua-non without which no hydrographic survey can be conducted - or indeed any survey of the coastline and the inter-tidal zone - is a knowledge of the tidal regime along the coast and across the waters concerned. In areas of complicated topography the tidal regime, both tidal height and range - the vertical element - and the times of high and low water - the movement of the tidal wave along the coast, the horizontal or temporal element - are correspondingly complicated. There may be seasonal variations superimposed on the basic 28-day lunar cycle. There will certainly be local meteorological effects which, in cases where the natural tidal range is small, may swamp the purely astronomical tides. So a series of sets of tidal observations, at a very large number of sites along the coast and, where banks offshore are concerned, there too, are needed, and for at least some of the tidal stations these observations will need to extend over a period of years. Of course, most developed countries already have a network of tidal stations and hence accurately established chart datums and tidal predictions. This is not, however, true of many less-developed parts of the world, where a cursory glance through Volumes II and III of the Admiralty Tide Tables, dealing with the world outside Western Europe and the Mediterranean, will show how little we know about much of the world.

Once the tidal regime, and hence the position of chart datum throughout the state's waters has been determined, we can proceed to the business of surveying and then drawing up the official charts on which the claimed baselines and limits are to be shown.

Since we are not concerned with the way in which the waters are to be used, we can for the moment forget the bathymetry below the low water line, except in so far as it is necessary to establish that drying line.

Instead of the laborious sounding over drying banks in small boats, much of the intertidal area, and thus the effective low water line, can be established by aerial photography or even, in the case of very remote areas, by satellite photography or other remote sensing. The photography or whatever needs to be tied to simultaneous tidal observations, or at least to accurate predictions, so that the correct "height" of low water can be established.

Unfortunately, in many areas where the position of the low water line is most variable and most critical this method breaks down. Some of the most productive oil and gas fields lie to seaward of the deltas of the great rivers, where the mud and sediments in the water make the penetration by any form of remote sensing virtually nil, and yet where the seabed slopes only gently and where small changes in height can shift the low-water line, and hence the outer edges of the territorial sea and the EEZ, a long way. There is still no substitute here for traditional echo-sounder surveys, though of course modern electronic position fixing systems and data processing computers make them a much less laborious proceeding than even ten years ago.

It is here, in the collation and presentation of the data, that modern techniques can and do help. And it is here that help is most urgently needed. To delineate, describe, and then to show on charts and as lists of coordinates, the baselines from which the breadth of the territorial sea is measured; and then provide the briefs from which the lawyers can start to argue about median or equidistant or equitably-dividing lines where adjacent countries' waters meet, for the countries of the world is an enormous task. No country has anything like the detailed knowledge of its waters and its claimed sea bed that it has of its land mass. Comparatively few have the organization or the qualified surveyors to set about collecting the information needed. There is, I hope I have shown, a massive task in geomensuration (I almost said Geomancy). It is with some relief that I turn over to Dr Francis to deal with Geophysics.

The EEZ, the Continental Shelf and modern surveying techniques

A.S. LAUGHTON.
Institute of Oceanographic Sciences
(Read by Dr. T.J.G. Francis.)

Admiral Morris has discussed the technical problems faced by hydrographic surveyors in determining the landward boundaries of the EEZ and has touched on some of the broader problems that fall outside conventional surveying. In particular he has discussed the determination of the baselines from which the boundaries of the territorial sea and hence most of the other boundaries are determined.

I want to consider the determination of the outer boundaries of the areas of national jurisdiction, giving special consideration to the coastal states with broad continental shelves where they can claim extension beyond the 200 nm EEZ under Article 76 of the UN Convention.

But first I must emphasise that EEZs may be proclaimed unilaterally outside the UN Convention. By 1984, 59 coastal states had made such declarations including the USA in March 1983. Most of these declarations claim rights similar to those that would be achieved under the UN Convention and extend to 200 nm. The discussion of modern surveying techniques to support these claims is relevant only so far as precise navigational standards are required to determine the distances or geographical coordinates by which they are defined. Clearly the new Global Position Satellite System, GPS, will play a significant role here.

I will confine my remarks to the EEZ and the Continental Shelf (CS) as defined by the 1982 UN Convention. The latter category arises from the inclusion in the 1982 Convention of the rights given in the 1958 Convention on the Continental Shelf - the Geneva Convention.

You will have heard from previous speakers the rights of coastal states to the resources, both living and non-living, that can be exploited in these two regions. Whereas in the EEZ these rights include living and non-living resources of the sea bed, the subsoil and the superjacent waters, including fishing rights and rights to produce energy from the water, currents and wind, in the area of the Continental Shelf beyond the EEZ, these rights are limited to the resources of the sea bed and subsoil.

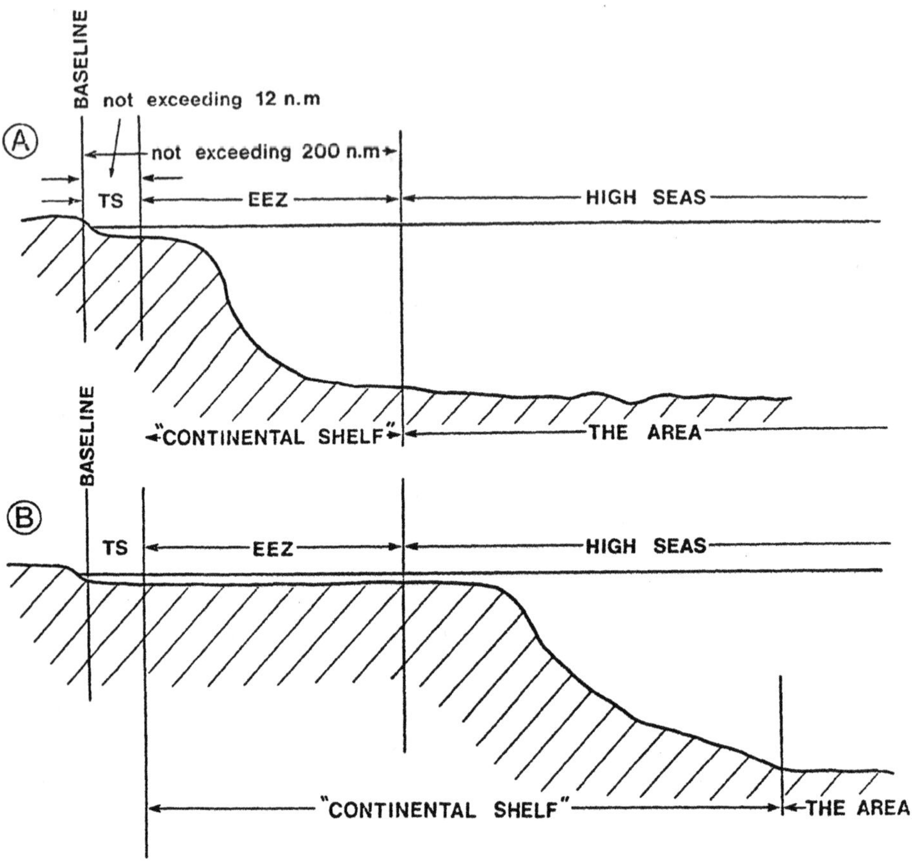

TS = TERRITORIAL SEA

EEZ = EXCLUSIVE ECONOMIC ZONE

Fig. 1. Definition of geographic terms used in the Convention. The "Territorial Sea", the "Exclusive Economic Zone" and the "High Seas" refer to areas at sea level. The "Continental Shelf" and the "Area" refer to the seabed and the subsoil.

(A) the case where the outer edge of the geological continental margin does not extend beyond 200 n.m. (Art. 76 para 4(a)). The legal continental shelf then extends to the same limits as the EEZ.

(B) the case where the outer edge of the margin extends beyond 200 n.m. and is defined by either the foot of the slope plus the 1% sediment line (Art. 76 4(a) (i)) or by the foot of the slope plus 60 n.m. (Art. 76 4(a)(ii)).

It is important that the terms used in the Convention are both defined and understood. Fig 1 shows the two cases (a) where the continental shelf, as generally understood by geologists, does not extend beyond 200 nm and (b) where it clearly does. The term "EEZ" applies to the sea and the land beneath the sea within the zone whereas the term "Continental Shelf" and "The Area" apply to the seabed and subsoil only.

To exercise their rights and duties the coastal states must define their boundaries rigorously and unambiguously and justify their claims.

There are three sorts of boundaries to these areas, the landward boundary, the boundary with neighbouring or opposite states and the seaward boundary bordering on the international zone. All of these depend on the baselines which have been described by Admiral Morris.

The landward boundary to the EEZ is the outer limit of the Territorial Sea as defined in Articles 3 and 4.

The boundary with neighbouring or opposite states is one that should be resolved by negotiation or if necessary by legal arbitration (Articles 74 and 83). The arguments may include a wide variety of factors such as precedent, equitable principles, geology, geomorphology, equidistance, etc. There is a growing body of case law on delimitation issues some of which has depended heavily on the geological or structural continuity that requires a knowledge of regional geology. In most places in the shelf seas and oceans this is inadequately known.

It is the seaward or outer boundary, however, which requires most detailed consideration here.

The outer boundary or limit of the EEZ is relatively easy to determine being 200 nm from the baselines from which the Territorial Sea is measured.
This involves standard surveying or geodetic techniques.

The outer limit of the Continental Shelf, however, is a complex matter determined partly by the shape of the sea bottom, partly by depth, partly by distance and partly by thickness of sediment. The formulation of the many criteria in Article 76 was the result of compromise between interested parties in the Law of the Sea Conference and has many ambiguities. It will provide a field day for international lawyers.

I quote:-

"The Continental shelf ..... comprises the seabed and subsoil of the submarine areas that extend beyond its territorial sea throughout the natural prolongation of its land territory to the outer edge of the continental margin, ..." and "the continental margin .... consists of the seabed and subsoil of the shelf, the slope and the rise". (Note that the shelf is defined as comprising the shelf, the slope and the rise!!)

Where the outer edge of the legal Continental Shelf - (not to be confused with the geomorphological continental shelf which geologists understand to end at the shelf break at about 200 metres depth), extends beyond 200 nm, the boundary shall be straight lines not more than 60 nm in length joining points conforming to the following criteria:-

(a) the thickness of the sedimentary rocks at the points shall be at least 1% of the shortest distance from such point to the foot of the continental slope,

<u>or</u> (b) the points shall not be more than 50 nm from the foot of the continental slope,

<u>also</u>(c) the points shall either not be more than 350 nm from the baseline, or not more than 100 nm from the 2500 m isobath <u>except that</u> on submarine ridges outside the continental margin the 2500m depth plus 100nm criterion will not apply.

(d) In the absence of evidence to the contrary, the foot of the continental slope is the point of maximum change of gradient at its base.

Fig 2 shows how the outer limit, based on the 1% sediment thickness criterion, can be applied in a case where this gives a limit beyond the 60 nm. Both criteria depend firstly on the important line, <u>the "foot of the continental slope"</u> as defined in (d) above. Anyone familiar with bathymetric profiles of the continental slope will realise that on any one profile there may be many changes of gradient especially where it crosses submarine canyons.

Clearly some kind of averaging of many profiles will be required to determine a continuous line. Complete bathymetric surveys, corrected for variations in the velocity of sound in seawater, and for the geometrical effects of side echoes will be necessary. However, even with these it may not be possible to draw a continuous line which represents the maximum change of gradient.

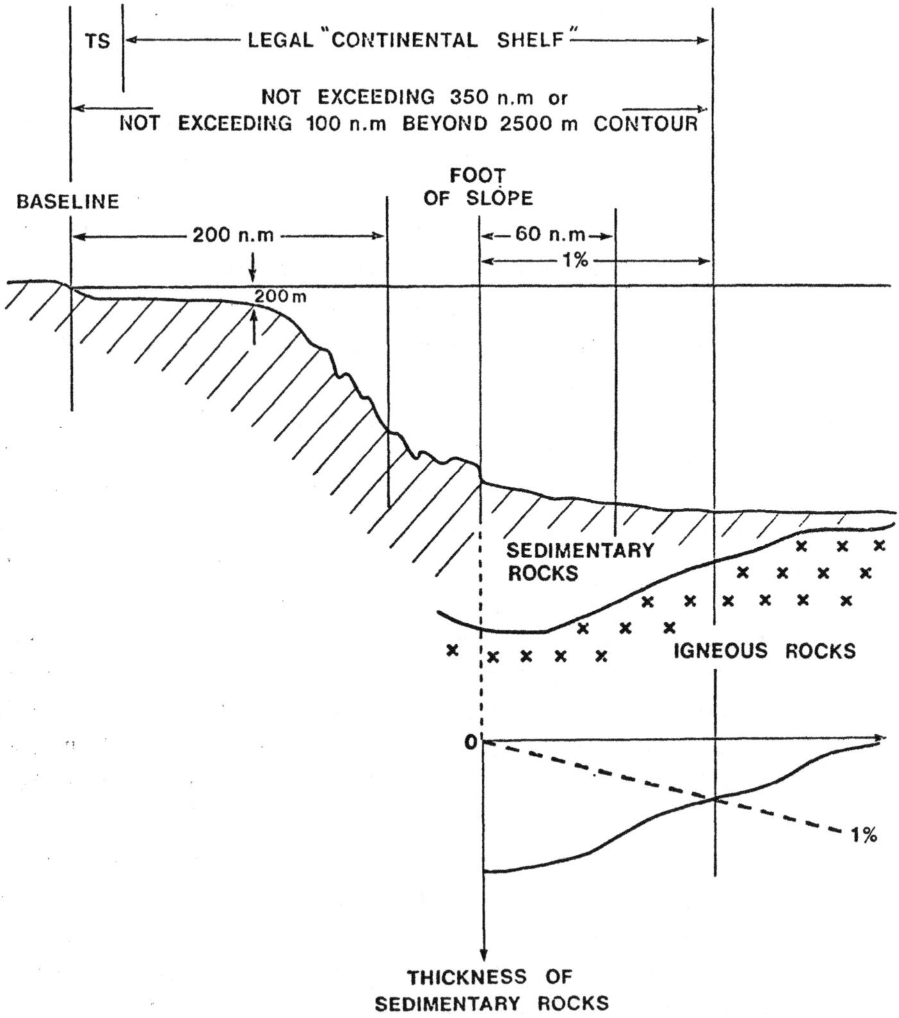

Fig. 2. The determination of the outer limit of the continental shelf according to Article 76 para 4(a)(i). The limit is drawn where "the thickness of sedimentary rocks is at least 1 per cent of the shortest distance from such a point to the foot of the continental slope".

Suppose the maximum change of gradient climbs up the slope or is in the middle of the slope. At what depth is it disqualified as not being "at its base"?

Having determined a foot of the slope line, constructing a line 60 nm from it is no problem. Remembering that the points to establish the outer boundary of the CS can be 60 nm apart, discontinuities or small embayments in the Foot of the Slope line such as might occur at the mouth of a canyon will be smoothed out.

So far the only field techniques that are required to be used are the conventional ones of bathymetric surveying, although there are significant interpretational problems.

However an alternative option in Article 76 is to determine points where the sediment thickness is at least 1% of the distance to the foot of the slope. Here seismic profiling methods will be necessary, similar to those used for oil prospecting. These profiles require expert analysis using velocity data to derive thicknesses of sediment layers. The identification of the reflecting horizon representing basement may be difficult. Even the meaning of the word basement is sometimes ambiguous since it may be the subsided edge of the rifted continental crust, itself comprising sedimentary rocks, or it may be the basaltic rocks of oceanic basement. To clarify ambiguities and to verify interpretations it may be necessary to drill holes down to basement in water depths of several thousand metres. These are technically possible, as has been shown in the Deep Sea Drilling Project, but may be expensive.

Article 76 places overall constraints on the extension seawards of the Continental Shelf. The outer edge of the Shelf shall either not be more than 350 nm from the baseline, or not more than 100 nm from the 2500m isobath. These do not present problems once the bathymetry is well established. However geological criteria once again arise when the 2500m + 100 nm criterion is excluded in Article 76 in the case of "submarine ridges". This exclusion however "does not apply to submarine elevations that are natural components of the continental margin, such as its plateaux, rises, caps, banks and spurs", in contrast to submarine ridges such as the mid-ocean ridge. The distinction between these features is not clear cut and will have to be based on a knowledge of the geology and a careful definition of terms.

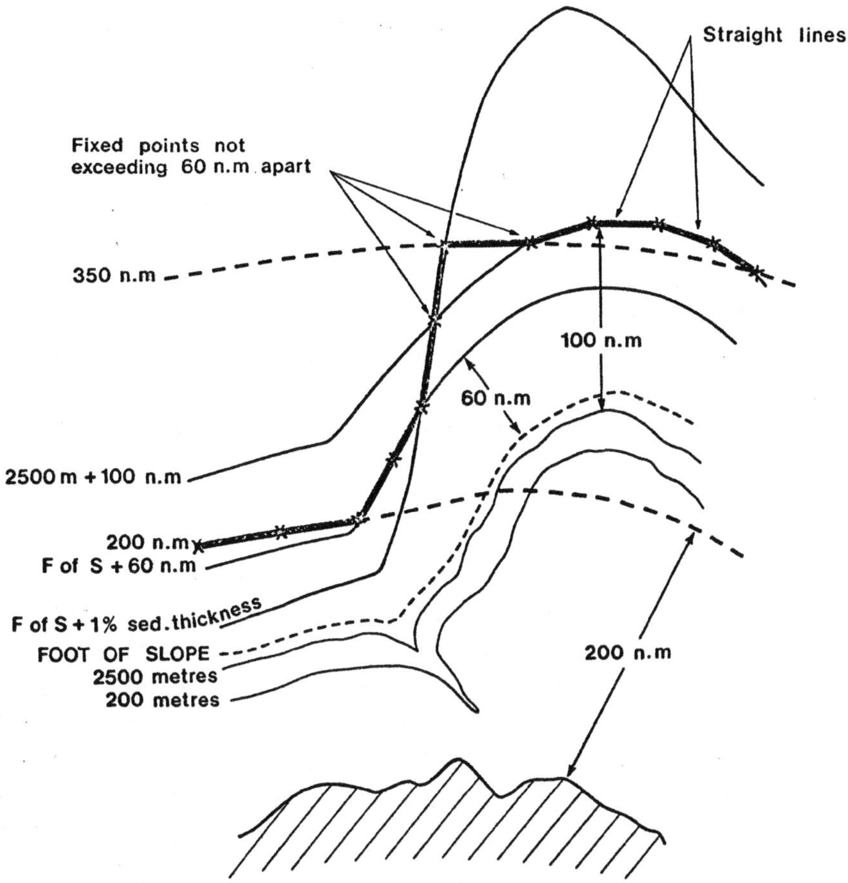

Fig. 3. An illustration of the outer limit of the continental shelf (thick line) in which it is determined by the various criteria in Article 76. From left to right these are:-

(a) 200 n.m. (para 4(a))
(b) Foot of slope plus 60 n.m. (para 4(a)(ii))
(c) Foot of slope plus 1% sediment thickness (para 4(a)(i))
(d) As (c) but truncated by 350 n.m. (para 5)
(e) As (c) but truncated by 2500 m plus 100 n.m. (para 5)

Note that the exclusion of extension beyond 350 n.m. on submarine ridges (para 6) is not taken to apply here since the submarine elevation illustrated is a natural component of the continental margin.

Thus on a typical broad continental margin a number of criteria can be used to determine the outer boundary of the legal Continental Shelf as are shown in Fig 3. The boundary jumps from one criterion to another as the morphology changes along the margin.

The onus is on the coastal state to submit its case for an extension of its Continental Shelf beyond the EEZ to a Commission of the UN on the Limits of the Continental Shelf (the so-called Boundary Commission). This submission must be made within 10 years of the entry into force of this Convention for that State. The limits "established by a Coastal State on the basis of these recommendations shall be final and binding". This commission shall comprise experts in the fields of geology, geophysics or hydrography, but not, interestingly enough, law.

I want now to consider the surveying requirements for the resources in the EEZ or the CS.

It will be clear that in all cases the EEZ or the CS will extend beyond the shallow geological continental shelf and will include a high proportion of water of oceanic depths. Already we are familiar with the hydrocarbon resources of the shelf. Since the continental slope is a prolongation of continental geology it is probable that hydrocarbons will also underly the slope in places. The exploitation may be difficult and expensive but the resources are almost certainly there. The Coastal State will want to know in detail about the environmental conditions in which the exploitation can take place, and will in particular want to know about the stability of the sedimentary slopes on which they will be operating. There is extensive evidence of slumping on the continental slope on small and on huge scales although little is known about how frequent these slumps are or whether they were a feature only of the lowered sea level during the last glaciation.

Other resources of the deep ocean, such as manganese nodules, hydrothermal sulphides or cobalt rich manganese encrustations occur in areas that are characterised by the texture or morphology of the sea floor. Coastal States will want to assess the potential resources lying in their EEZ or CS as soon as possible since this will determine the strength with which they argue for delimitation with their neighbours or with which they push their CS claims seaward, bearing in mind the 10 year time limit for making these claims.

Between 35 and 40% of the oceans may in future lie within the jurisdiction of coastal states, as EEZ or CS. It is clearly a Herculean task to undertake bathymetric surveys of these areas with the degree of completeness needed to define potential resource areas. Other methods of rapid reconnaissance of the sea floor shape are needed to indicate the critical areas for further detailed surveying.

Side scan sonar can be used as a first tool. The long range system, GLORIA, sweeping a swath 20 nm wide at 9 knots can cover 4 thousand sq. miles a day giving an acoustic picture of the morphology of the ocean floor similar to the Landsat pictures from satellites. You will hear later from Gary Hill of the US Geological Survey, how the US is using GLORIA to make a reconnaissance survey of its EEZ.

Such a survey must then be followed up by detailed mapping, preferably using multi beam swath echo-sounders, to delineate seamounts, active and extinct sea floor spreading ridges, fold mountains, slump features, etc. Near-bottom studies with manned or autonomous vehicles, detailed geophysical studies and geological sampling will follow as resources are progressively evaluated.

No one supposes that these resources will all be identified, assessed or exploited in the near future. Supply and demand, national wealth, world economics and technology will all have to play their part in the timing of these activities. But in this decade a unique and irreversible step is going to be made in the ownership of resources from the sea which will place considerable demands on marine technology and science in meeting the political and economic needs of coastal states.

LIVING RESOURCES, ASSESSMENTS AND QUOTAS

by

R Jones

DAFS Marine Laboratory
Aberdeen

Abstract

This contribution deals with factors affecting productivity in the sea and methods of determining the size of fish resources. Fish production varies considerably throughout the world oceans and in EEZ's, largely due to variations in primary productivity and the variations in the length of the food chain. Methods of resource estimation include the area swept method, the method of "virtual population analysis", egg survey methods, and the use of acoustic methods.

Details are given of the principles on which long-term and short-term management advice is given. Long-term considerations usually lead to the concept of an optimum harvesting strategy; for example some combination of fishing effort and mesh size that is expected to maximise yields in the long-term, irrespective of the absolute values of these yields. Short-term considerations are usually concerned with the determination of quotas to meet particular objectives; for example to maintain the current level of fishing effort, to reduce fishing effort, or to increase stock size.

Finally, details are given of the procedure by which scientific advice is translated into management action.

Introduction

With growing pressures on fish stocks, and the development of EEZ's, many fish stocks throughout the world have come under increasingly strict management, and the need for advice, based on reliable resource assessment has become more and more important. This paper is concerned with the technical aspects of fish stock assessment, with means of assessing sustainable productivity, and with means of making short-term predictions of catches, and recommendations about quotas, to meet a range of management objectives.

Levels of Production

Whilst it is important to know the size of a resource when it is about to be exploited, it is even more important to know the long-term sustainable level of production that might be expected from that resources. In general, long-term sustainable fish yields are closely dependent on the level of primary production and on the number of steps in the marine food chain from plants to man. These are both variable entities and some details and examples are given below, starting with a comparison of productivity on land with productivity in the sea.

On land, there is essentially a very short foodchain from plants to man. Typically this consists of plants - herbivores - man.

In the sea, foodchains are usually longer than this for the simple reason that marine herbivores are very much smaller than their terrestrial counterparts so that more steps are required to get from primary producers (ie plants) to animals such as a herring or cod that are large enough to be harvested economically.

The difference in size of the herbivores is typically very large indeed . For example a typical territrial herbivore is a sheep or a cow, whereas in the sea it is likely to be a copepod about the size of a grain of rice.

At each stage in the food chain, about 85-90% of the food input is lost in respiration, so that the relatively longer marine foodchain may appear to have a potentially lower productivity to man than the terrestrial food chain. This limitation is largely compensated for however by the fact that most of the plant production on land is not grazed but goes into the detrital cycle. In the sea on the other hand, about half of the spring bloom and about 90% or more of the summer production is grazed.

In the sea, there can be considerable variation in productivity as illustrated by the three examples in Figure 1.

The highest levels of productivity are found in the so called upwelling areas of the world. These are characterised by hydrographic regimes, where there is a continuous or intermittent upward transport of nutrient rich water to the euphotic zone.

This has the effect of sustaining a higher level of productivity, than would be possible if there was only a recycling of nutrient within the water column. Major upwelling areas are found along the west coasts of North and South America, off the west coasts of Africa and South Africa, and in the Antarctic. In these regions, plant species can include not only large unicellular organisms, but also species that form chains up to several millimetres in length that are large enough to be eaten by fish such as anchovy and sardine. One important upwelling area is off the coast of Peru, where an important species, the anchovy, depends directly on plants at some times of the year but on zooplankton such as copepods at other times of the year. There are therefore one to two steps from primary production to man. Potential production in upwelling areas can be very large and for a time in the 1970s the Peruvian anchovy fishery for example, yielded about 10 million tonnes per year.

The next most important kind of production area is found on the continental shelves of the principle land masses where there are typically two to three steps from primary production to man (Fig. 2b). Here, typical plants consist of unicellular organisms about 100 μ in diameter which are grazed by copepods, which in turn are grazed by various primary carnivores including fish species. Man typically exploits primary carnivores such as herring and mackerel, or secondary carnivores such as cod, hake and other whitefish species. This kind of food chain is typical of that found in many EEZ's.

The lowest levels of productivity are found in the deep ocean where there are typically three to four steps from primary production to man. Factors that tend to limit productivity in the deep ocean are:

a) great depth and the presence of a persistant or semi-persistent thermocline which limits the rate of upward transport of nutrient from deep water.

b) there is relatively little exchange with nutrient richer adjacent water masses.

c) the production is largely depended on recycling of nutrients within the euphotic zone. Associated with this, typical plants tend to be much smaller (about 25 μ in diameter) than elsewhere.

d) many of the primary carnivores are themselves small invertebrate species in the zooplankton so that a great deal of the secondary production is used up before it reaches even small fish.

The net result is that 3-4 steps are required to get from the primary producers to animals large enough to be worth harvesting by man.

These differences in productivity, and the fact that the deep oceans are difficult and potentially uneconomic to exploit, has led to the principle fisheries being concentrated on the continental shelves of the major land masses and in Antarctica.

<u>The Relationship Between Stock Size and Productivity</u>

Although it is sustainable productivity in the long-term that is important it is also necessary to know something about the size of a resource and the relationship of this with productivity.

Figure 2 for example shows that a stock is influenced by 4 principle factors. There are two factors that cause a stock to decline ie these are the yield, due to fishing (Y) and natural losses (M) due to predation and natural causes.

On the plus side, a stock is typically augmented each year by an annual influx of juveniles normally referred to as "recruits" (R). There is also an increase in stock size due to the growth of individual animals (G). In the long-term, therefore if a stock is neither to collapse nor explode, the following relationship should hold.

$Y + M = R + G$

or, rearranging terms

$Y = R + G - M$

This is a useful relationship as it draws attention to three important factors that influence yield, ie recruitment, individual growth and natural losses.

Effect of Fishing

An analysis of the effect of fishing on a stock is somewhat simplified by the fact that only the terms G and M in equation (1) are directly effected by fishing. Although one might expect recruitment to be influenced by stock size (which in turn is influenced by fishing) observations show that for most of the commercially important teleost fish species recruitment is independent of stock size over a very wide range of stock sizes (Fig. 3). Because of this it is customary to make a preliminary analysis of the effect of fishing on a stock in terms of the effect of fishing on growth and natural losses.

Figure 4 shows a generalised relationship between sustainable yield and stock size. As stock size increases, sustainable yield is expected to increase but only up to a point. However large a stock may become, the maximum sustainable yield can only depend on the maximum possible level of productivity and theoretical considerations frequently suggest that this is expected at some intermediate level of stock size as indicated in the diagram.

Sustainable Yield and Fishing Effort

Since fishing tends to decrease stock size, Figure 4 can also represent the relationship between sustainable yield and fishing effort if one considers the horizontal axis with opposite sign and change of scale. The result doing this is shown in Figure 5. Sustainable yield is expected to increase with increasing fishing effort up to a certain point but further increases in fishing effort are expected to cause a decline in sustainable yield. Theoretically there is some maximum sustainable yield (MSY) at some intermediate level of fishing effort.

Natural Fluctuations

Figures 4 and 5 show an idealised state of affairs for a hypothetical resource that receives a constant number of recruits each year. In practice however, fluctuations in recruitment can be large in the short-term, and cyclical in the long-term. As a result even with constant fishing effort a succession of annual yields is usually very variable as indicated in Figure 6.

Optimum Fishing Strategy

When the existence of natural fluctuations in recruitment is combined with simple fishing theory one arrives at the concept of an optimum fishing strategy. This is indicated in Figure 7, which shows 2 times series of yields. First, the solid line, represents the sequence of annual yields that is expected from the fishery if fishing effort is kept constant.

Secondly, the dotted line shows what is expected to happen to these yields if, at time t_1, there is some change in fishing strategy that is expected to increase potential yields in the long-term. There might for example be a reduction in fishing effort or an increase in mesh size. In either case, the immediate effect of would be a decrease in the yield at time t_1. This would be followed by a transition period, until time t_2 when the expected long-term benefit would start to take effect. The dotted line then shows how from then on, it is expected that annual yields would be larger than they otherwise would have been. Due to the inability of scientists at present, to predict future levels of recruitment, it is not possible to predict such a future sequence of yields in absolute terms. What can be predicted however is the extent to which future yields should be greater (or less) than they otherwise would have been for any given change in fishing strategy.

Economic Considerations

If one were dealing with a resource that was under the ownership of a single company, management policy would probably be largely influenced by long-term considerations as described above. A likely objective would be to maximise yields or more probably to maximise profits or other economic benefits in the long-term. In practice however marine resources are typically exploited by large numbers of independent fishermen. So long as it is profitable to do so, new entrants continue to come into the fishery and fishing effort increases and tends to a level in excess of that required for generating the maximum sustainable yield or of maximum total profit. Figure 8 shows the principle economic consideration from the point of view of a single boat owner. As total fishing effort increase, the average size of the resource declines and gross revenue per boat declines. Running costs per boat will not be affected by total effort to the same extent however, and eventually a level of effort will be reached at which revenue per boat = running cost per boat. At this level, profitability will be zero, and there will be no incentive to build more boats.

Short-term Considerations

In practice, for the reasons indicated above, many stocks are exploited at levels well above those required for maximising retainable yields or profits. This maybe acceptable to the industry, provided there is no danger of a collapse of the stock, due to recruitment failure. In this situation short term considerations may become important, and advice is often sought about measures suitable for preventing any further increase in effort, or decline in the resource.

Figure 9 shows the principle factors involved in making a short-term prediction. If one considers a stock in 1987 for example this will consist partly of the survivors of the individuals that were alive in 1986. It will also contain a number of recruits that have been added to the population during the intervening year. A short-term prediction therefore involves a consideration of:

what is the stock size in 1986

how many of these individuals will survive to 1987

how much will the weight of these survivors increase due to growth of the individual

what will be the size of the recruiting yearclass in 1987.

The basic factors are exactly the same as those involved in long-term theoretical considerations, except that it is also necessary to know something about stock size in absolute terms.

Resource Estimation

There are four principle methods used for estimating the absolute size of a resource. These are the areas swept method, the virtual population analyses method, the egg survey method, and the use of acoustic surveys.

Area Swept Method

This method is based on the catches of fish per unit fishing time using a net with known catching characteristics, fished over a known area of sea bottom. The method is appropriate for fish that live close to the sea bottom and is frequently used in countries that lack the resources for carrying out any of the other methods. A limitation of the method is that it is usually difficult to determine what proportion of the fish in the path of the net are actually retained. However a range of estimates can be obtained for a probable range of catching efficiencies.

Virtual Population Analysis

This is essentially a theoretical procedure that sets out to calculate the number of fish that must have been present in the sea in order to account for the number of fish that are known to have been captured by fishermen. The method is particularly suitable for long-lived individuals that can be aged and particularly for those that are relatively heavily exploited and for which there is a reliable sampling programme to provide estimates of the numbers caught.

Egg Production Method

For short-lived species, annual recruitment may make up a relatively large proportion of next years catch, so that an analysis of historical data, using the virtual population method, may not be very useful. For species with pelagic eggs, the size of a resource, or at least of the spawning biomass of females, may be determined by estimating the total egg production for a season. For example at high latitudes, many fish produce a single spawning batch each year during a fairly predictable spawning period. An estimate of egg production can then be determined by sampling eggs with a research vessel at the appropriate time of the year. Spawning biomass can then be determined from the relationship

number of female spawners = number of eggs produced/average fecundity per female

In sub-tropical waters there may be multiple spawnings and it is not such a straightforward business to determine either the number of spawnings per year or the number of eggs produced per spawning. Recently however a method has been developed for getting round this difficulty, based on histological observations on post-ovulatory follicles.

Acoustic Surveys

Another method that can be particularly useful for short-lived species, involves the use of acoustic techniques. Acoustic surveys provide a means of determining the biomass of those species that are capable of reflecting an acoustic signal, so that in the first instance, this kind of survey provides an estimate of what is known as acoustic biomass. To determine the biomass of any particular fish species, it is also necessary to carry out an associated net sampling programme to determine the probable species composition of the various acoustic traces. One of the difficulties with this method is the interpretation of acoustic signals due to the difficulty of determining the target strength of different sizes and species of fish. Research on this problem is still continuing.

Comparison of the Acoustic Surveys and the Egg Production Method

The acoustic method and the egg production method each have certain advantages and disadvantages for stock assessment. The acoustic method for example gives a wide coverage, but not necessarily for only one species. It can be carried out at relatively short notice, it includes both juveniles and adults, and the results can be worked up relatively quickly. It provides information about the distribution and relative abundance of species but there can be considerable uncertainty, about the absolute values of stock biomass.

The egg production method on the other hand, is so far only applicable to a few species, it provides only information about the spawing individuals in the population, and takes a relatively long time to work up the results. It is likely to be more reliable than the acoustic method for providing absolute estimates of stock size however. The two methods are frequently regarded as complementary rather than alternatives.

Estimating Recruitment

Because recruitment to exploited fish stocks may represent an important part of each year's catch it is an important in any short-term prediction to be able to provide a reasonably reliable estimate of the number of recruits expected each year. This is normally achieved by carrying out a research vessel survey, to determine the catches per unit fishing effort of members of a new year class when they are large enough to be retained by special small mesh nets, but still small enough not yet to have entered the fishery. These catches provide a set of recruitment indices which can be plotted against subsequent recruitment as measured in later years by other methods. Once this has been done it is possible to conduct a survey to determine a recruitment index from the catches per unit fishing time of juveniles and then, to convert this directly to an estimate of the number of fish that are expected to recruit to the stock in the following year.

Scientific Working Groups

Under the present management regime, stock size predictions, and recommendations about appropriate catch levels for appropriate objectives are required each year. Normally this leads to the following procedure for each stock of commercial importance.

Scientists from those countries having an interest in the stock concerned meet at some central venue, nowadays usually in Copenhagen. They put together their respective national data for the stock concerned, and make short-term predictions along the lines indicated above. It is customary to make predictions to satisfy various objectives, such as a) to ensure that stock size next year will be no lower than it is at present, or b) to ensure that stock size next year will have increased by some stated amount

Estimates of future catches needed to satisfy stated objectives are usually known as TAC's (Total Allowable Catch).

The reports of these various working groups are passed to a body known as the ICES Advisory Committee on Fishery Management (ACFM) which summarises the findings of the working groups in a form suitable for reading by administrators and fishermen. This information is passed to the fishery administrators of each country who subsequently meet and negotiate to agree subdivision of the TAC's into National Quotas for each of the important stocks. In practice, negotiation may sometimes lead to national quota allocations which, when summed, differ from the scientifically determined TAC.

Relevant parts of the ACFM Report are also passed to the Northeast Atlantic Fisheries Commission (NEAFC), the International Baltic Sea Fishery Commission (IBSFC) and the North Atlantic Salmon Conservation Organisation (NASCO). They are also made available to other management bodies, especially the individual governments of ICES member countries and to the Commission of the European Community (EC). In the case of EC countries, the ACFM advice is first scrutinised by the Scientific and Technical Committee for Fisheries (STCF) and this report provides the Commission with a basis for forming recommendations to member states and ultimately to the Council of Ministers.

The ACFM report serves two purposes. It reviews the state of the stocks and provides options for management, but it also provides a record of the information and arguments underlying the advice. The report is therefore detailed and voluminous which can create problems in the vital step of conveying the results to the people most directly affected, ie the industry. We have tried to overcome this in the UK by periodic seminars with fishermen and also by the publication of a digest of stock information in the Fish Stock Record and Fishing Prospects and also periodic reports in Fishing News.

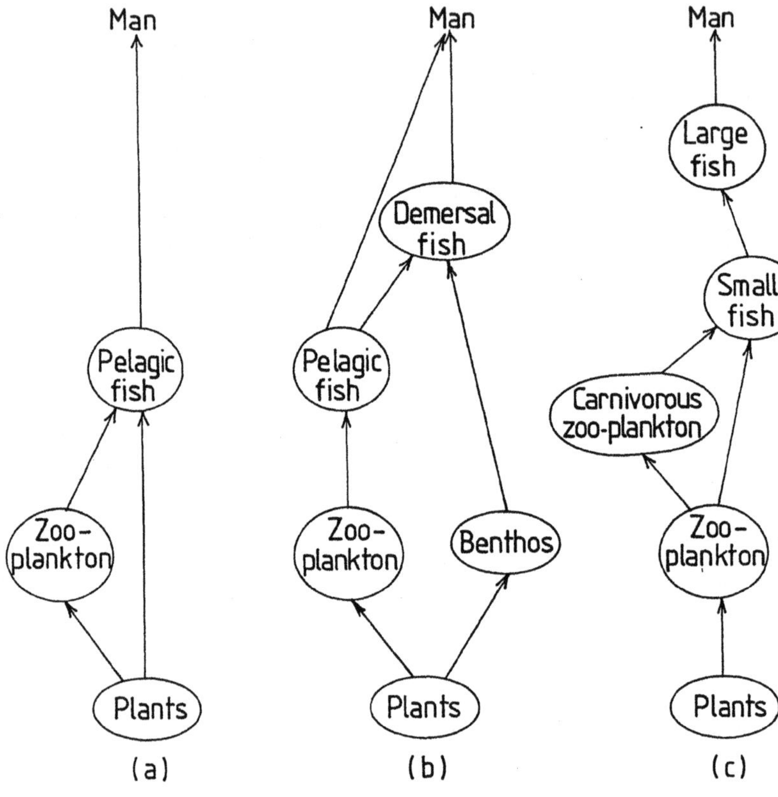

Figure 1 Food chains from three different kinds of ecosystems

a) An upwelling area

b) A Continental Shelf area

c) Deep ocean

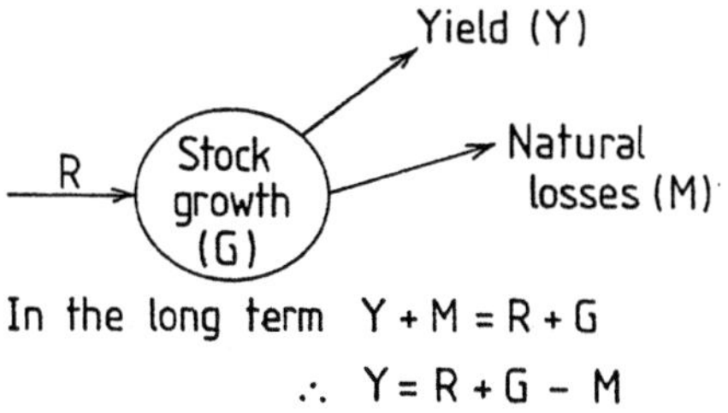

Figure 2 Showing the principle factors influencing the size of a resource in the long term.

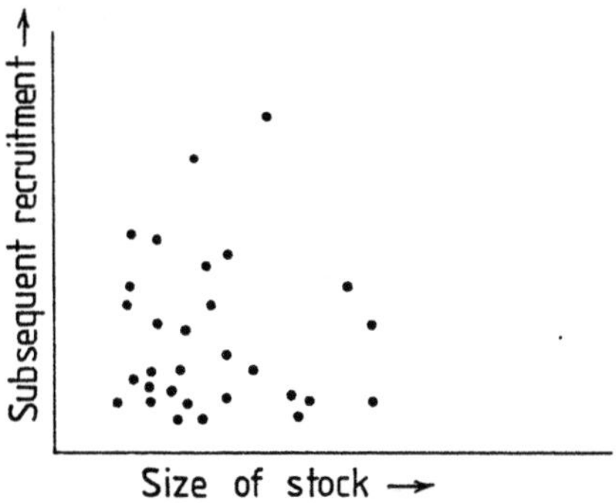

Figure 3 Kind of relationship often found between stock size and subsequent recruitment (from Jones, R. 1973. Density Dependent Regulation of the numbers of cod and haddock. Rapp. Proc. Verb. Cons. int. Explor. Mer., 164: 156-173).

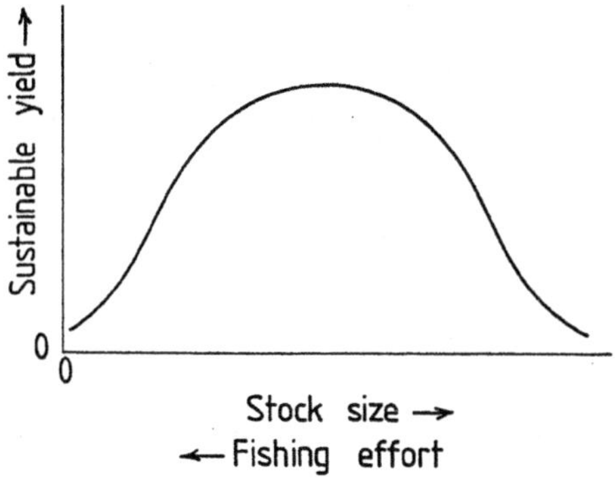

Figure 4 Theoretical relationship between sustainable yield and stock size.

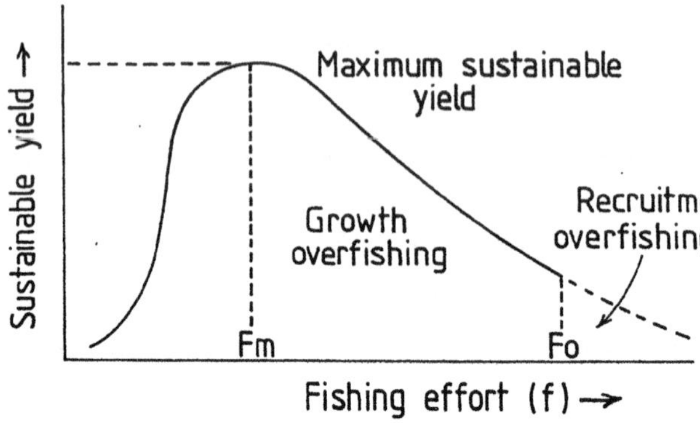

Figure 5 Theoretical relationship between sustainable yield and fishing effort.

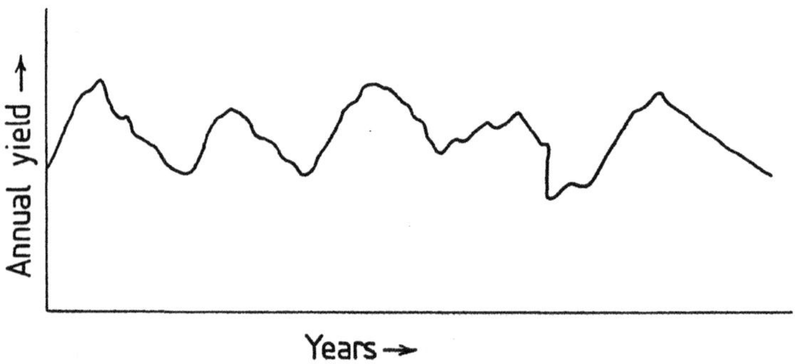

Figure 6 Showing how in practice, annual yields tend to fluctuate considerably due to natural variations in recruitment.

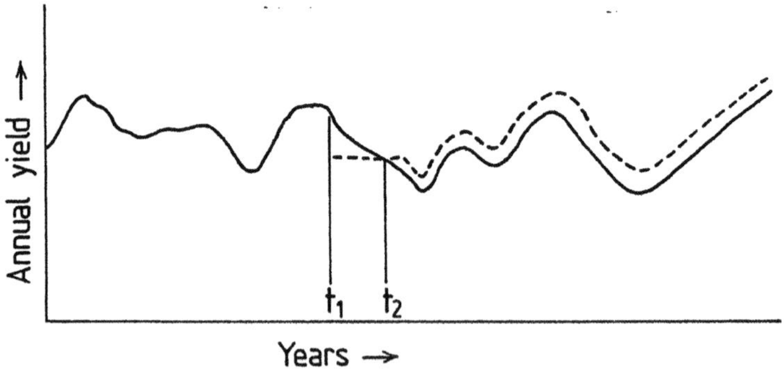

Figure 7 Showing the effect on annual yields of a change in exploitation strategy at time t_1. The solid line shows annual yields without change in effort. The dotted line shows annual yields if exploitation strategy is changed at time t_1.

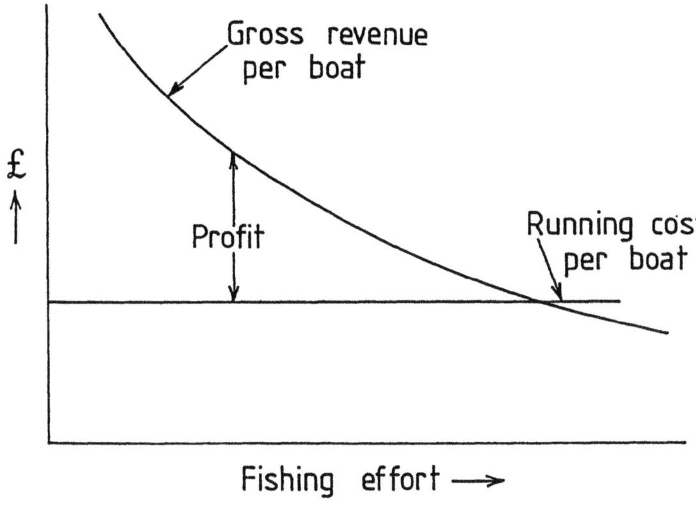

Figure 8 Economic considerations. Effect of total effort on profitability from the viewpoint of a single boat.

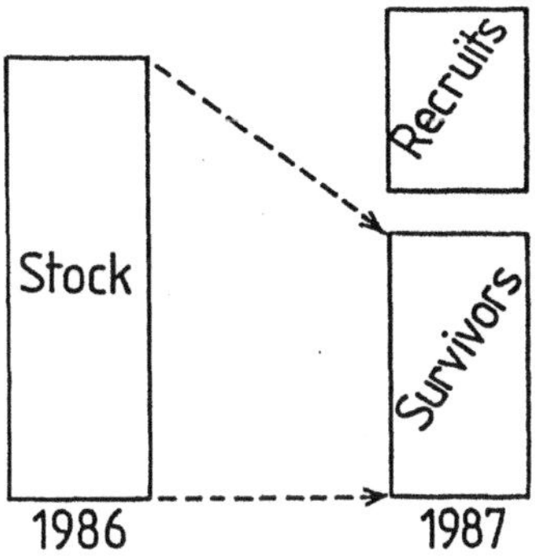

Figure 9 Principle factors concerned in a short term prediction. Stock in 1987 consists of survivors from stock in 1986 plus fish that recruit to the stock in 1987.

Opportunities and Management of EEZs

The scale of EEZs with particular reference to areas of US and UK juristiction: problems associated with exploitation and also protection of EEZ resources.

A view of the UK position by Professor A J Smith

Department of Geology, RHB New College, University of London

Abstract
The UK Government has not declared an EEZ around the UK or any of its dependent territories claiming that it does not see the need to claim such zones in the knowledge that the UK and majority of dependent territories already have 200 mile exclusive fishing zones and, as a matter of international law, rights to their continental shelves which are inherent and do not have to be proclaimed.

Such a view sadly reflects official attitudes to Britain's maritime role: it contrasts with the Presidential Proclamation by President Reagan on 10 March 1983 which led to the exciting developments to be described elsewhere in these proceedings. Worse, the view of even the informed public regarding our national knowledge of the UK equivalent of an EEZ seems to be that the oil companies know what they need to know and that otherwise little is known of this vast and immediate area. In fact the UK continental shelf is perhaps the best surveyed and understood of any such shelf in the world - starting with surveys by academic institutions, the main thrust has been taken up by the British Geological Survey and, to-date, an enormous compilation of data covering sediments to seismicity in map and computer form is available to anybody who seeks that knowledge. To do this work new methods of instrumentation were developed and worked from a variety of specialist vessels. The creation of an EEZ would however, to the north west of the UK, take us beyond the continental shelf into exciting deeper conditions where future resources await development. It is in these deeper conditions that the UK could demonstrate its widest capabilities to best effect.

Further, the creation of EEZs around our remaining dependencies - Anguilla, Bermuda, British Indian Ocean Territory, British Virgin Islands, British Antarctic Territory (particularly the South Shetland and South Orkney Islands), Cayman Islands, Falkland Islands (and South Georgia and South Sandwich Islands), Gibraltar, Hong Kong, Montserrat, the Pitcairn Group, St Helena (and Ascension and Tristan da Cunha) and Turks and Caicos Islands - brings in a total of about 2,000,000 square nautical miles of sea and ocean. Indeed, depending upon interpretation, the area could be somewhat larger. In this vast area the geological regimes included every type of situation - continental shelves and slopes, abyssal plains, ocean ridges and ocean trenches - each with its many exciting scientific and economic potentials.

By establishing EEZs for the UK and its dependent territories major opportunities open for the United Kingdom. The nation can, for this vast area, prepare an inventory of resources, demonstrate its interest in the oceans, increase public awareness of the importance of the seas and oceans, encourage the development of technology, show the capabilities of British industry, aid the dependencies and offer a focus for British manufacturing and surveying abilities thus utilising the considerable existing national expertise. Such a demonstration will undoubtedly lead to greater opportunities in the rest of the world, starting with the Commonwealth of Nations.

Introduction

The UK Government has not declared an EEZ around the United Kingdom or any of its dependent territories claiming that it does not see the need to claim such zones in the knowledge that the United Kingdom and the majority of dependent territories already have 200 mile wide exclusive fishing zones and, as a matter of international law, rights to their continental shelves which are inherent and do not have to be proclaimed. It must be at once admitted that to declare such EEZs would raise sensitive issues in relation to the UN Law of the Sea Convention of 1982. Britain has not signed that Convention though as this is written, April 1986, 159 states and other entities have and 26 have ratified. Furthermore, on the basis of Article 305 of the Convention, those dependencies which enjoy full internal self-government and which have competence over the matters governed by the Convention can sign the Convention for themselves. Thus it might be argued that the United Kingdom should not interfere with that freedom of some of its dependent territories.

Being an island state on a geologically passive margin, the United Kingdom has a reasonably large adjacent continental shelf. The largest part of this shelf is already divided along agreed lines - particularly the North Sea - based on the equal distance principle. The UK could have pressed for a larger area in the northern North Sea - ie. as far as the Norwegian Trough. This claim was not pressed twenty five years ago because few, perhaps nobody, envisaged the presence of major oil fields in the northern North Sea. To the west and south of the United Kingdom the equal distance principle severely limits the area that can be claimed. Indeed to the south west of the United Kingdom our area is almost 'squeezed-out' by the Irish and French claims. Only to the north west of the United Kingdom does the area extend further out and there much remains to be settled between the conflicting claims of Ireland, Denmark, Iceland and Britain.

Notwithstanding the position taken by the Government of the United Kingdom and the nature of conflicting claims, what do we know about that area which we call the UK continental shelf? All too often I hear, and this from people who should know better, the view expressed that little is known. This view is based partly on the impression that only the oil companies have, through their contractors, undertaken surveys and that the Hydrographer to the Navy is on public record in expressing his concern about the slow rate of progress at producing new charts of the waters around these islands. What is missed is the tremendous achievements of the British Geological Survey's Marine Surveys Directorate staffed by some 140 persons using British ships and, often, specially developed equipment. Let us consider that Directorate's achievements.

The state of knowledge of the UK continental shelf

Since 1967 the British Geological Survey (BGS), and since the mid 1970's contracted by the Department of Energy, has been charged with (1) the preparation of a series of reconnaissance maps at a scale of 1 : 250,000 of the continental shelf and, as appropriate, the slope to show solid geology, the Quaternary geology and sediment distribution with the additional collection of gravity and magnetic data, (2) the assessment of hydrocarbon resources to provide an independent catalogue of those resources and reserves and (3) the acquisition of all geological and geophysical data emanating from the UK continental shelf in order to maintain a national archive for any purpose which may be in the national interest.

A vast fund of sea floor and subsurface data has been accumulated. The Marine Surveys Directorate oversees marine geological and geophysical research groups. It has more than 220,000 nautical miles of geophysical records, many at a 10 km grid interval (Figure 1). Over three quarters of these records include gravity, magnetics and seismic profiles - including Boomer, Pinger, Airgun or Sparker and side scan. In more recent years techniques have been developed to run all types of seismic operations at the same time providing the best data for all forms of interpretation. The development of a deep-towed boomer, capable of operating at water depths of up to 1 km, now permits the recovery of extremely high quality data which reveals bedding relationships with great precision. The geophysical data are backed by more than 25,000 cores up to 6 m length - a sample about every 10 km of track. Such cores have been taken anywhere on the shelf, indeed vibrocored samples can be taken at water depths up to 1½ km and drilled samples at depths to 500 m. Vibrocoring or rotary drilling are performed from the same frame deployed on a 2.5 km umbilical. Operations to depths of 2 km are possible for vibrocoring and will shortly be for coring. In addition, there are over 500 long cores drilled at sites of special geological interest using a purpose designed wireline drilling system, some of them using dynamically positioned drilling vessels (Figure 2). All these data are held on computer with an interactive data-base which allows immediate plotting of results. All information is held on an international exchange format.

All the data which have been obtained are openly available: the maps (see Figures 3, 4, 5, 6, 7, 8, 9) are well advanced (about ½ of possible coverage already), the remainder in the next 5 years. Such progress is in contrast to much complained about progress in land surveys. Maps show a very wide range of data and interpretation. The sea floor depths are presented in a contoured form, with a disclaimer that they are not to be used for navigation together with sediment-type; Quaternary sediments (ie. laid down in last 2 million years and therefore of great significance for offshore installations) are displayed with details of geotechnical properties and the solid, ie. pre-Quaternary, geology is shown. Indeed in many maps careful assessment and interpretation of geophysical and borehole data has permitted a sub-set of maps revealing the geology and geological structure at progressively deeper and thus older horizons. To my knowledge no other country has prepared such a series of maps. Not yet mentioned is the study of earthquakes in the UK offshore areas: some of these are of noteworthy intensity with epicentres in the very heart of the oil producing region. Such a study is of both immediate importance and critical for understanding the passed and continuing evolution of the shelf region (Figure 10).

Equally important, though confidential, is the work of the Hydrocarbon Research Group. These geologists, by law, receive all data obtained by commercial surveys. It is their task to make entirely independent assessments of the hydrocarbon resources from fields currently identified, being exploited, or awaiting exploration, exploitation or leasing. In a few sentences in the Department of Energy's Annual Brown Book this group's findings are on view. Those few words belie a great wealth of information which can help in formulating national policy and ensuring the development of offshore hydrocarbons in a pattern to be in the best national interest.

Such work is only possible from considerable application by experts: it is the first task in the protection of any EEZ. What has been and is being done thus contrasts with views often expressed in 'informed' quarters that the UK shelf is barely explored. That hydrographic surveys have not moved fast enough is, sadly, a reflection of availability of funding and thus of equipment and purpose: this said let it not be doubted that the UK shelf is well surveyed geologically. That the term 'reconnaissance' is used may cause some misunderstanding yet the scale is about "One mile represented by a quarter inch" and many a motorist is happy enough to set off into unknown districts with maps of this scale or less. It is not my remit to talk about living resources: other contributors will apply themselves to this, but so far as non-living resources are concerned, a great deal has been done. What is lacking is a public awareness of what has been and is being achieved in the continental shelf adjacent to the UK landmass. Moreover, in many quarters, there is uncertainty about what to do next.

The need for a programme

Such a picture is in sad contrast to reports emerging from other countries, particularly Japan, France and the United States. In the last, the Proclamation of the President of the United States of America on 10 March 1983 that the area from a line three nautical miles off the coast of the United States and its island territories was the Exclusive Economic Zone of that nation. As described elsewhere in these proceedings this gave the United States juristiction over the living and non-living resources in 3.9 billion acres. This vast acreage includes the area contiguous to the 48 states, the states of Alaska, including the Aleution Islands, and of Hawaii and the Hawaiian chain, together with the U.S. island territories in the Caribbean and those spread across the Pacific Ocean, many of the last mandated territories governed by the USA since World War II. This Proclamation was not inhibited by the nicety that the United States of America, like the United Kingdom, was not a signatory of the Convention. Its declaration, however, has electrified that nation into an undertaking to prepare an inventory of its EEZ. Within this U.S. Zone nearly every style of geobathymetric regime occurs. This alone, at first sight, even for the cognisenti, makes those in the United Kingdom concerned with maritime affairs envious. It is against this background that I return to the desirability of an announcement of EEZs for the UK and its dependencies. Even something less than a declaration will suffice: an intention to investigate will go much of the way.

If we take the area immediately adjacent to the United Kingdom (see Figure 11), such an announcement would embrace a large area of passive continental margin to the north west of our islands with an extension down into depths greater than 2 km. Though small this area has considerable economic potential and some remarkable features which point to a better understanding of how margins evolve and how such margins might be subject to submarine sliding on a major scale. It is an area in which a range of techniques can be developed, some of which already exist and have already been used in the slope region. A programme of detailed work in this region would be a useful beginning. Even if it were undertaken on the basis of no prejudice to other claims, it would set UK marine science and technology a target, a focus, and an opportunity for achievement. Such a programme is, however, not enough. Britain needs to look to the world ocean: to do this Britain should look to its dependencies.

The world ocean and its floor: the target

Before discussing a plan for the United Kingdom, a brief description of the potential resources and principal geological features of the world ocean is required. In part, these have been discussed earlier in the proceedings, but there remains a need to list them here.

a) The possible uses and resources of adjacent maritime zones include:

i. Non-living:

a. Aggregates, mineral placers, hydrocarbons, coal, water, ores, sulphur, phosphates, polymetallic nodules and crusts, polymetallic sulphides.

b. Disposal sites of non-toxic, toxic and radionuclide wastes.

c. Man-made structures, viz platforms, artificial islands, power generators (wind, wave, tide, current, OTEC).

d. Extraction of elements from sea water, and desalination.

ii. Living:

a. Food from fish and crustacea.

b. Resource base for biotechnology (from flora and fauna).

iii. Employment and Recreation: - including recognition of sites of scenic, scientific and historic value.

b) The principal features:

i. Continental Shelf - a relatively flat sea floor out to the shelf break. In legal parlance it extends to the 200 m isobath but, in fact, the depth of the shelf break depends on a number of geological factors. The width of a continental shelf is also dependent on geological factors.

ii. Shelf break - the point where the slope steepens between the shelf and the continental slope.

iii. Continental Slope - a steeper slope (up to 1 in 15) which might be cut by submarine canyons and affected by submarine landslides. Extends from the break to the continental rise.

iv. Continental Rise - a less steep zone characterised by the accumulation of sediment derived from the land mass, shelf and slope. The depth of change from slope to rise will depend on many factors but in general terms this is about 2 km. The rise extends out to the abyssal plain.

v. The Abyssal Plain is a reasonably flat area. This may be interrupted by abyssal hills and ridges and escarpments.

vi. Abyssal Hills - may be extinct submerged volcanoes or features associated with ocean ridges.

vii. Escarpments - features which usually lie at right angles to the ocean ridges. Sites of transform faults.

viii. Ocean Ridges - a broad feature which flanks the crests where ocean floor is created. The depth of the crest ranges from about 2000 metres to being above sea level. The crests are sites of active processes where magma reaches the sea floor (or above the sea floor). The ridges are the biggest and most continuous features on the planet's surface.

ix. Ocean Trenches - the sites where the ocean floor on its lithospheric plate is consumed by moving beneath an adjacent lithospheric plate.

They are characterised by being much deeper than the ocean floor and are usually long and narrow.

x. Lithospheric Plates - the components of the surface and outer 70-100 km of the planet. The surface of the plates may be ocean floor or ocean floor and continent. The plates are being constructed along the crests of ridges and destroyed by subsidence in the vicinity of the trenches.

c) The potential resources associated with the features listed in b):

i. The shelf a. Non-living resources reflect the features defined, ie. shelves may have carbonate and siliclastic sediments, placer deposits of economic importance and phosphates. The sub-sea strata may contain coal, oil, gas, sulphur, water and various ore bodies. The nature of the resources will be, to a great extent, a reflection of the geological character of the adjacent land.

b. Living resources will include varieties of fish and crustacea as well as submarine plants. The types and abundance will depend on a range of physical factors (depth, light, temperature, salinity) and nutrient supply.

ii. The slope will resemble the shelf though the living resources will reflect the greater depth. Where deep ocean water upwells the waters may be rich in nutrients.

iii. The rise has a thick pile of sediment and may be a long-term target for hydrocarbons.

iv. The abyssal plain may be the site of polymetallic nodules (Mn, Fe, Co, Cu, Ni). The plains may offer the best sites for the disposal of medium and high level radionuclide wastes.

v. The abyssal hills and escarpments may be the site of the metallic crusts (Co, Fe, Mn).

vi. The ocean ridge and flanks of the ocean ridges can be expected to contain polymetallic sulphides of possibly greater value (though harder to recover) than nodules. Waters above some ridges should be richer than usual in nutrients.

vii. The ocean waters. Biologically, these can be fertile, or veritable deserts with very little life. Much depends on water movements, especially in the vertical sense. Chemically, they contain many valuable elements including Uranium.

The United Kingdom's dependent territories

Britain currently is directly responsible for seventeen dependent territories (see Figure 12) set in the North Atlantic, the Caribbean, the South Atlantic, the South western Atlantic, the Indian and the Pacific Ocean.

What resources exist with the potential EEZs of these territories is a matter of conjecture: some guesses can be made and a better informed estimate might be possible by a more careful perusal of published information. I am particularly conscious that the living resources may be better assessed than I have indicated. In each case I have limited my account to an interpretation on the basis of an EEZ being 200 nautical miles from the coastline: I have not attempted to consider the other interpretations which are admissible.

1. <u>North Atlantic</u>

a) Bermuda (32°20'N, 64°45'W). An abyssal hill that was once a volcano now capped by a carbonate reef complex, the whole set in an abyssal plain which lies between the North American continent and mid-ocean ridge system of the North Atlantic Ocean. The size of the Zone is about 130,000 square nautical miles. The possible non-living resources within this zone include polymetallic nodules, though their quality is said to be inferior to those of the Pacific Ocean. I assume that there are living resources which need to be explored and there must certainly be a need to undertake 'conflict analysis', ie. the study of interaction between present and potential users: local people, tourism, carbonate extraction, fishing and, possibly, the extraction of yet to be discovered non-living resources.

b) Gibraltar (36°09'N, 5°21'W). Part of the Iberian Peninsula and any EEZ would be restricted by the conflicting claims of Spain and Morocco. Here there is, however, some potential for demonstrating the use of ocean current energy.

2. <u>Caribbean</u>

a) Anguilla, British Virgin Islands, Montserrat (approximately 18°00'N, 62°30'W). Part of the Lesser Antilles and situated near an area of collision between the western North Atlantic/ North American plate and the Caribbean plate. An area of great scientific interest and already the subject of much attention from British and American scientists. The combined potential EEZ includes shelf, slope, rise, trench and escarpments. The size of the zone is about 5,000 square nautical miles, the area being limited by the rights of other territories. The possible resources include minerals associated with plate boundaries, ocean and geothermal energy, hydrocarbons as well as living and recreational resources.

b) The Cayman Islands (19°30'N, 80°40'W). Carbonate topped islands, the base of which may be of igneous or sedimentary rocks. They have shelf, slope, rise and abyssal plain. The size of a 200 nautical mile zone is about 4,000 square nautical miles, the area being limited by the claims of Cuba, Mexico, Honduras, Jamaica and Colombia. Geologically an interesting area with much to be learned. The possible resources, with the exception of fish, are unknown.

c) Turks and Caicos Islands (21°45'N, 71°35'W). Carbonate topped islands, part of the Bahamas Bank, possibly on a piece of continental crust. They have shelf, slope, rise, trenches and abyssal plain within the 200 mile zone. Again it is already an area of considerable scientific interest. The size of the zone is about 5,000 square nautical miles, the area being limited by claims of Bahamas, Haiti and the Dominican Republic. The possible resources are carbonate rock, possibly hydrocarbons, ocean energy and living and recreational resources.

3. South Atlantic

a) St Helena (15°57'S, 5°42'W). A volcanic island on an escarpment about 400 nm east of the crest of the South Atlantic mid-ocean ridge and bordering the Angola abyssal plain. Has a small shelf, large slope and rise, an escarpment and abyssal plain with its 200 mile zone. The size of that zone is about 130000 square nautical miles. Possible resources are minerals, including polymetallic sulphides, and living resources.

b) Ascension (7°57'S, 14°22'W). A volcanic island close to the crest of the South Atlantic mid-ocean ridge. It has a small shelf, a slope, rise, mid-ocean ridge crest, ridge and escarpments. The size of its 200 mile zone is about 130,000 square nautical miles and the possible resources are polymetallic ores and fish.

c) Tristan da Cunha (37°15'S, 12°30'W). A group of islands (Tristan, Inaccessible, Nightingale and, 300 nm southeast of the main group, Gough) situated less than 200 nm east of the crest of the South Atlantic mid-ocean ridge. The islands have narrow shelves, then slopes, rises, ridge system, escarpments and abyssal plains. The size of the zone is about 250,000 square nautical miles and the possible resources are polymetallic ores and fish.

4. Southwestern Atlantic

a) The Falkland Islands (approx. 51°45'S, 59°00'W). Two main islands and many smaller islands. Continental in character and though part of the South American shelf, their geological affinity is with Africa. There is a wide continental shelf, slope and abyssal plain. Between Argentina and the islands there is probably an old, closed plate collision suture. The size of the zone exceeds 150,000 square nautical miles with the area limited by Argentinian claims - quite apart from Argentinian claims to the islands. The possible resources include minerals, hydrocarbons, coal and abundant marine life, the last already being reaped by Eastern Block and Iberian fleets. In addition these islands are of strategic importance for the development of the Antarctic continent.

b) South Georgia (54°30'S, 36°00'W). A volcanic island, with smaller 'rocks', situated on the Scotia Arc - the boundary between the South Atlantic/South American plate and the young Scotia plate. It has shelf, slope, an active trench and abyssal plains. The size of the zone is about 250,000 square nautical miles and the possible resources are minerals, hydrocarbons and marine life.

c) South Sandwich Islands (approx. 57°45'S, 26°30'W). A line of volcanic islands (Zavodovski, Visokoi, Candlemass, Saunders and Montagu) stretching along 120 nm of the Scotia Arc. They have shelf, slope, active trench and abyssal plains. The size of the zone is 180,000 square nautical miles. They, like South Georgia, are part of an area of great scientific interest and the possible resources are minerals, hydrocarbons and marine life.

Also included here, though regarded by some quarters as part of the Antarctic continent, are

d) the South Orkney Islands (60°35'S, 45°30'W).
(Coronation Island and Laurie Island).

and e) the South Shetland Islands (approx. 62°00'S, 60°00'W).
(Clarence Island, Elephant Island, King George Island, Livingstone Island, Deception Island and Smith Island).

Both d) and e) are of considerable scientific interest already, the islands being volcanic and associated with active geological processes at plate boundaries. Nearby there is a young ridge system developing and there is a considerable potential for polymetallic ores. Marine life appears to be abundant. The islands are separated from the Antarctic continent by the Bransfield Strait. The size of a 200 mile zone for d) and e) is about 250,000 square nautical miles, the area being limited by boundaries associated with the adjacent Antarctic landmass.

5. Indian Ocean

The British Indian Ocean Territory (7°00'S, 72°00'E). A group of islands, the Chagos Group, with the best known island being Diego Garcia. Carbonate topped, the volcanic islands are towards the southern end of the Laccadive-Chagos ridge, a feature perhaps related to a "hot-spot" in the planet and thus similar to the Hawaiian-Midway line of the Pacific Ocean. There is an extensive shallow shelf together with slope and abyssal plains. The flanks of the spreading Carlsberg Ridge are within 200 nautical miles of the archipelago. The size of a zone is about 200,000 square nautical miles and the possible resources are polymetallic ores and marine life.

6. Pacific Ocean

The Pitcairn Island Group (approx. 25°00'S, 120°00'W, and Oeno 100 nm to the NW, Henderson 120 nm to the NE and Ducie 250 nm to the E). The islands lie at the southeastern end of a chain of islands and abyssal hills and on the west flank of the East Pacific ridge. The islands are volcanic and have a carbonate cover. Each has a small shelf and an extensive slope; abyssal plains and hills, ridge flank and escarpments lie within their 200 nm zone. The size of the zone for the Group is about 270,000 square nautical miles and possible resources include polymetallic ores, mineral crusts, ocean energy and marine life.

The approximate total area of the zones described approaches 2,000,000 square nautical miles (about two-thirds of the area included in the U.S. Proclamation). It should be noted that not included in the foregoing description are the zones off Hong Kong, the Phoenix Group (near the Pitcairn Island Group), and Belize in Central America. Antartic continent claims are also omitted as is the Rockall extension. The last is, as has been indicated already, of special importance because of the scientific interest, hydrocarbon potential and proximity to the UK mainland.

The purpose and the benefits

British experience in maritime technology is considerable. It ranges from individuals with a fund of expertise, to special inventive skills, to laboratories with decades of tradition in marine biology and fisheries, to major accomplishments such as that of the Marine Surveys Directorate of the British Geological Survey for the contiguous shelf and the recognised abilities of the Institution of Oceanographic Sciences (why else should they be so deeply involved in the U.S. EEZ programme?), to the nation's industrial abilities shown in response to the needs of the offshore hydrocarbon industry be it in heavy engineering or services and supplies and the traditions of our fishing industry.

This experience and these skills can be put to global use as the World Community accepts the consequences of the implementation of the Convention of the Law of the Sea. Sadly, as the Sub-Committee of the House of Lords Select Committee concerned with Marine Science and Technology made clear in its report: British scientific and economic enterprise in this sphere lacks focus. Our position is seen to compare badly with the United States, France and Japan. If the Government of the United Kingdom saw fit to sign the Convention and thus declare an EEZ for the United Kingdom and, just as significantly, signed for or on behalf of its dependencies, then just such a focus would be created. By such an act the United Kingdom would declare its interest in the oceans and increase public awareness within the United Kingdom of the importance of that interest. By undertaking to "defend" those EEZs through exploration, there would be encouragement for scientific technological development as an inventory of the living and non-living resources was made; the dependencies would be considerably aided by such inventories and through new information HM Government would be better able to debate current and impending international diplomatic problems. Such work would actively demonstrate the capabilities of this country in the maritime scene and increase its chances of more work for Commonwealth and other countries. The people to undertake this work already exist.

Part of my brief was to talk about the protection of EEZs. I did not imagine it to include any requirement to talk about a military role. To defend something one must first define what one needs to defend; to defend it without recourse to force one must demonstrate an interest in or, better, an intimate knowledge of what you seek to defend.

No greater impetus to the considerable maritime abilities which exist in Britain is clear to me than the declaration of an Exclusive Economic Zone. The President of the United States of America in proclaiming that America would make claim to the contiguous Zone and that of its dependent territories and, more significantly, that an inventory of that Zone would be made, offered both a challenge and a focus. The United Kingdom, with its dependencies could offer a similar opportunity. An area of nearly 2,000,000 square nautical miles, the very breadth of which makes the challenge, both geographically and geologically, one which will be both stimulating and rewarding.

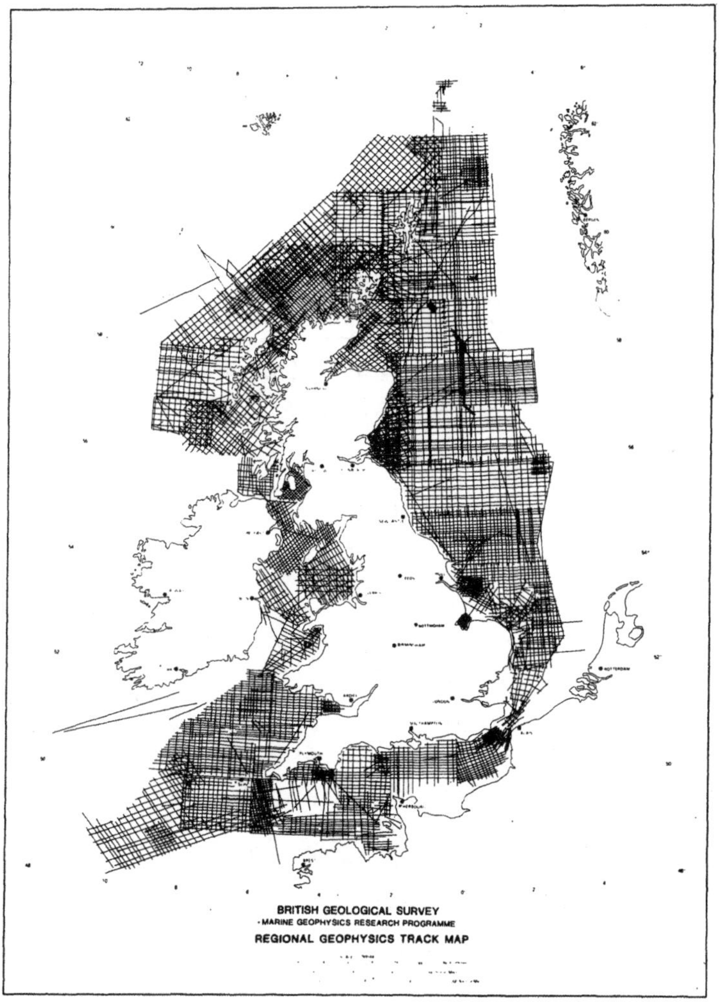

Figure 1. British Geological Survey - regional geophysics track map. (Figures 1 to 11 were prepared by the Marine Surveys Directorate and are reproduced here by kind permission of the Director of the British Geological Survey).

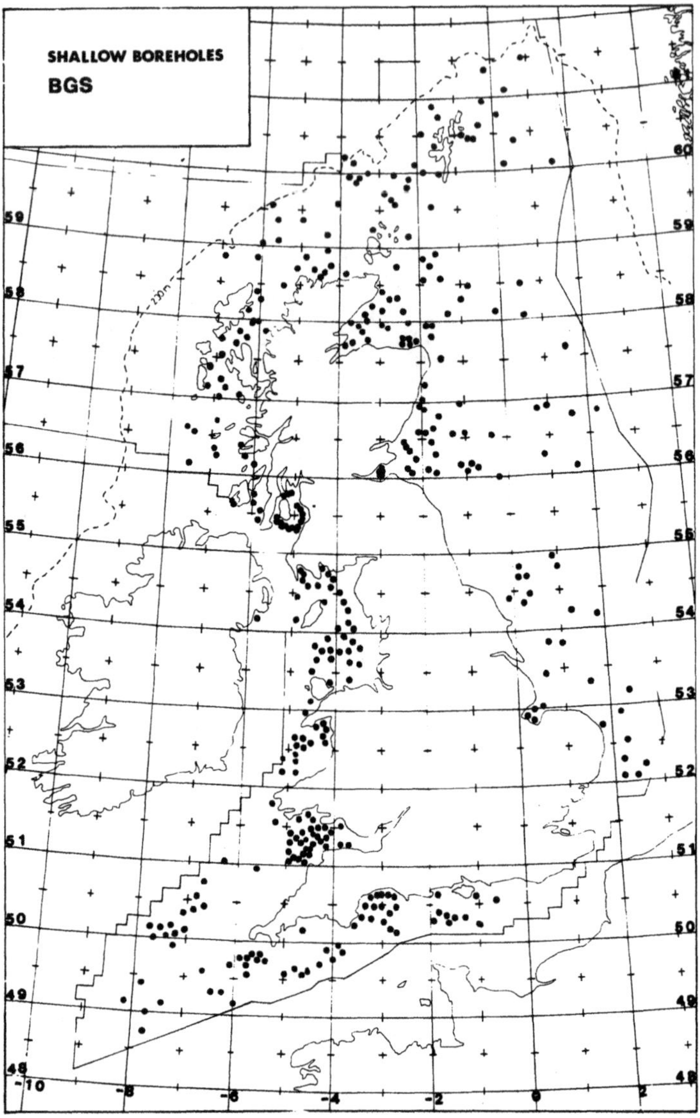

Figure 2. Sites of shallow boreholes made by the British Geological Survey (B.G.S.).

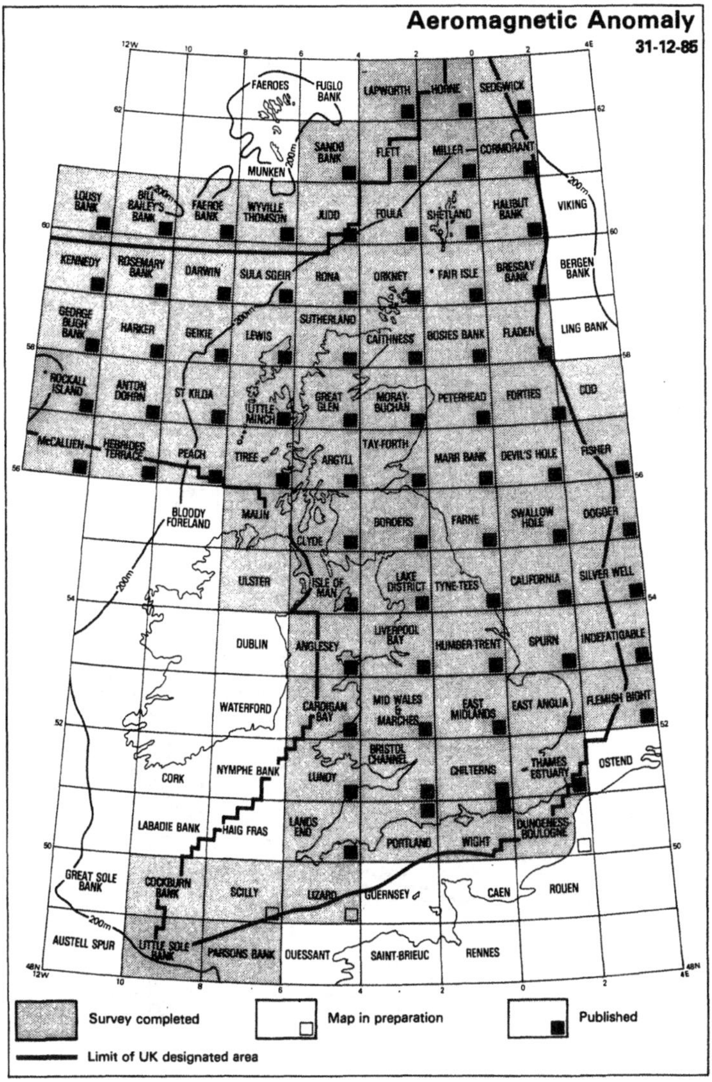

Figure 3. Aeromagnetic anomaly maps produced by the B.G.S. (All map sheets shown here and in figures 5 to 9 are published or are to be published at a scale of 1:250,000).

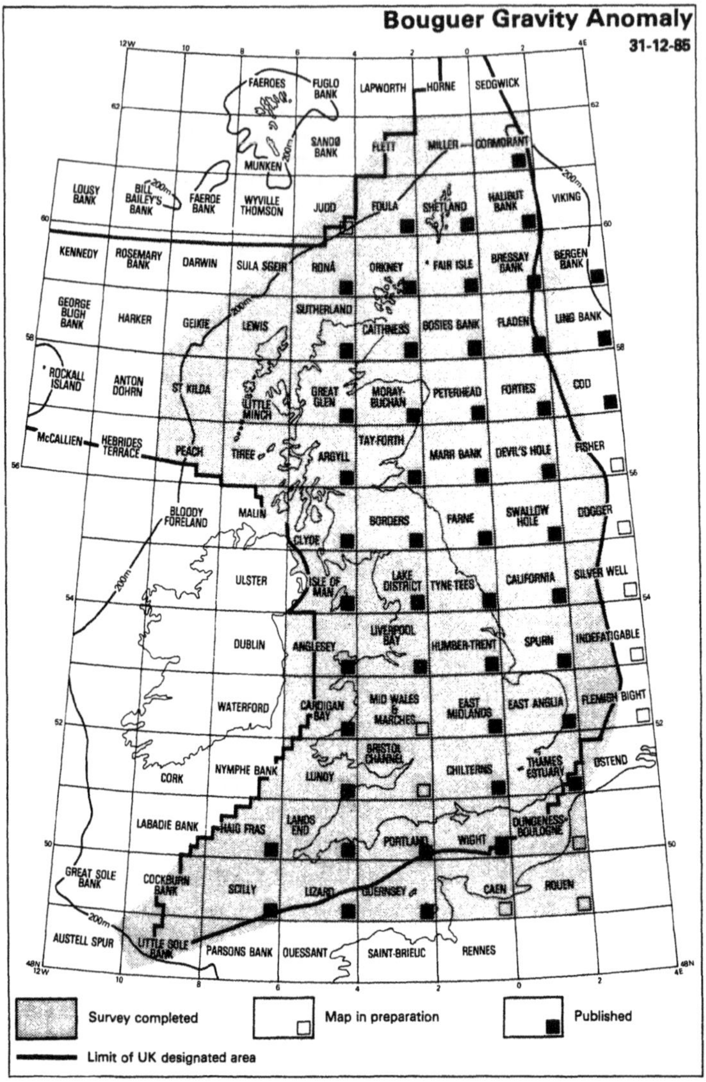

Figure 4. Bouger gravity anomaly maps produced by the B.G.S.

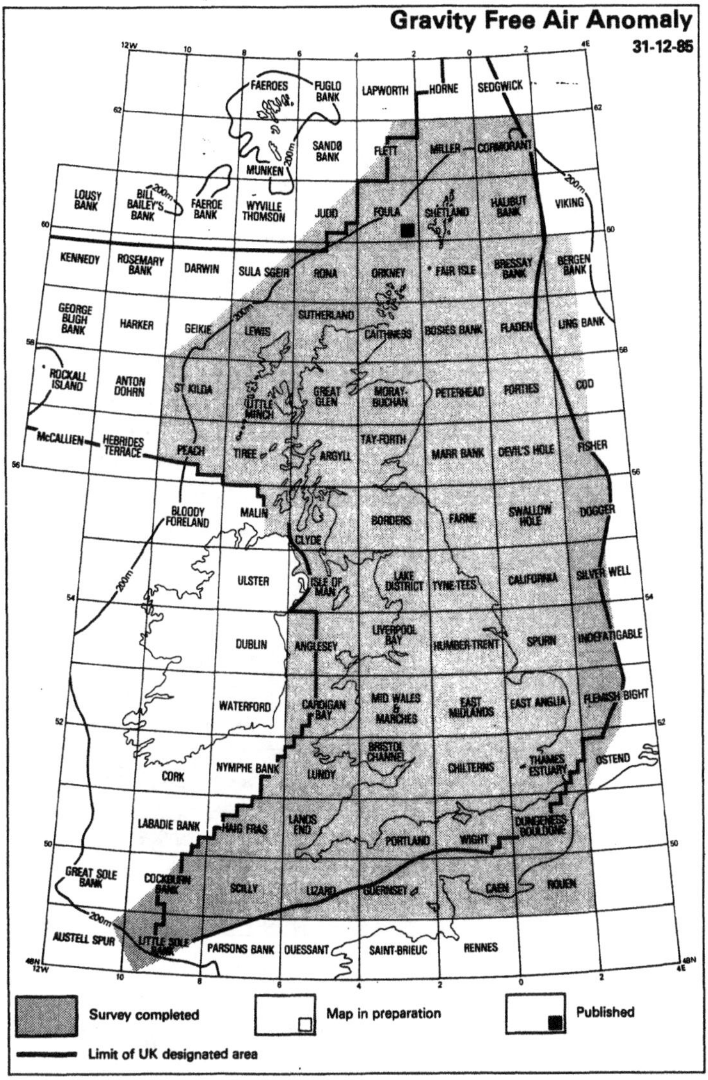

Figure 5. Extent of gravity free air anomaly surveys by the B.G.S.

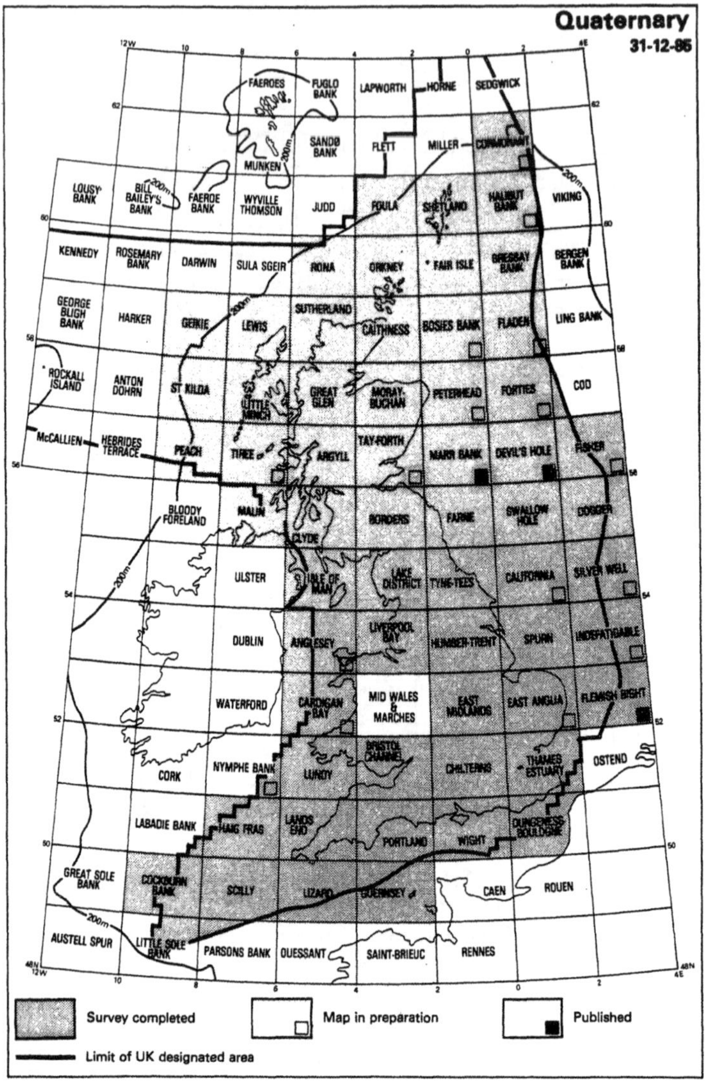

Figure 6. Extent of surveys of Quaternary geology with maps published and in preparation.

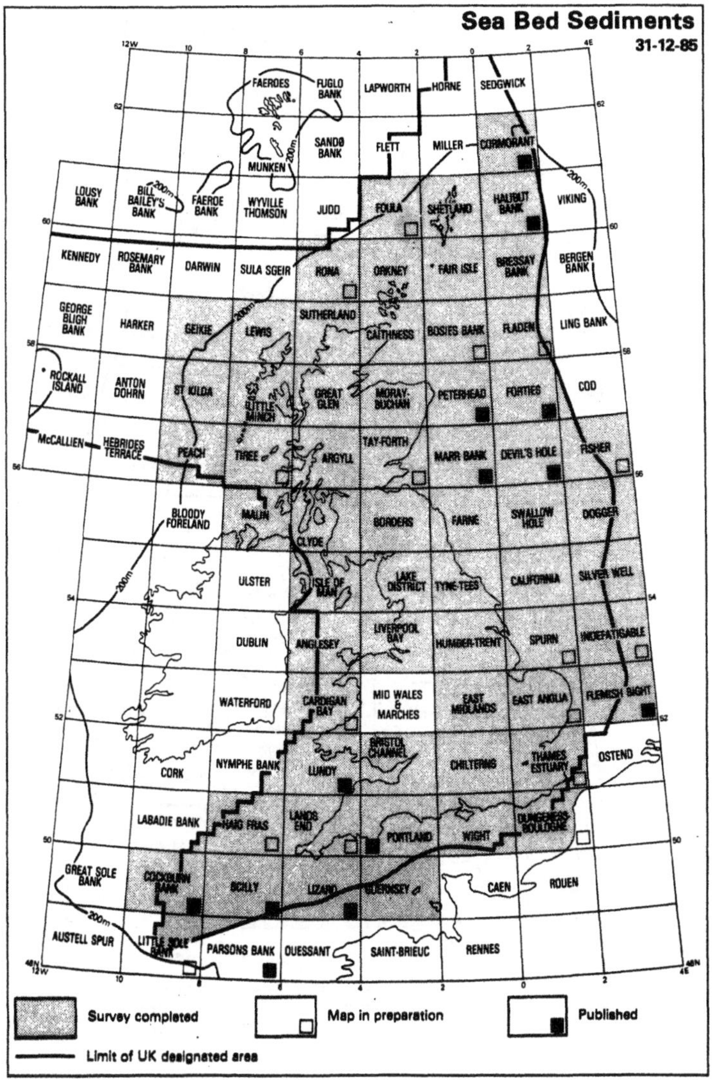

Figure 7. Extent of surveys of sea-bed sediments with maps published and in preparation.

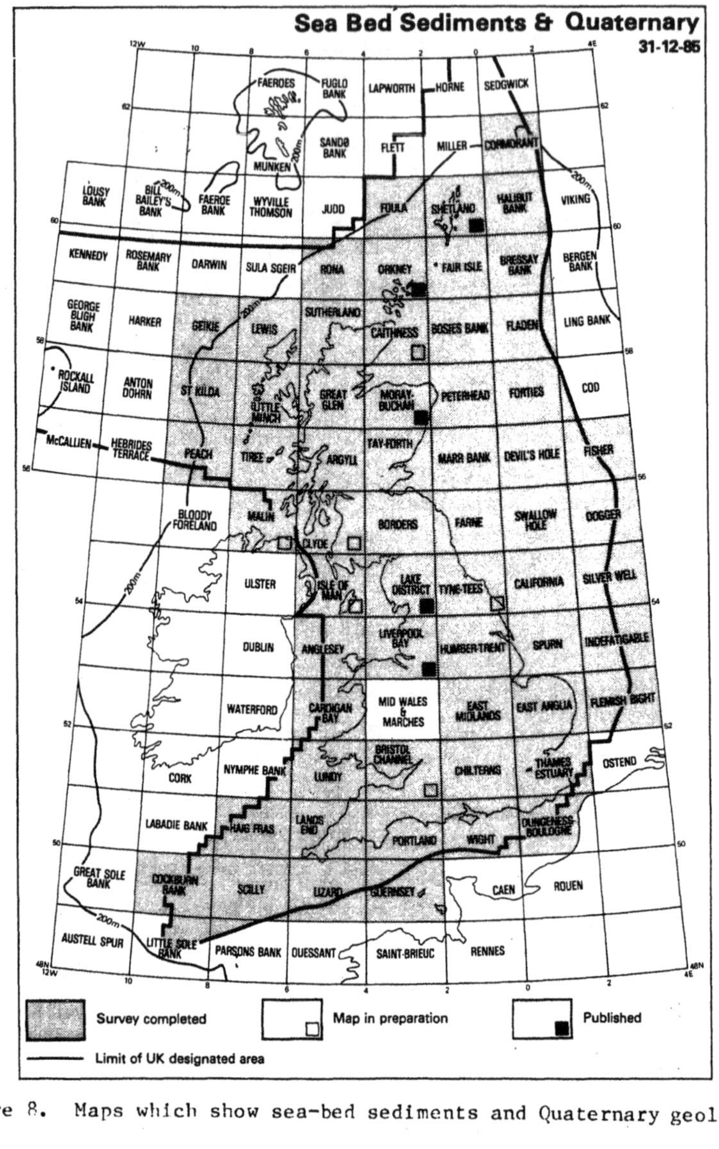

Figure 8. Maps which show sea-bed sediments and Quaternary geology published and in preparation.

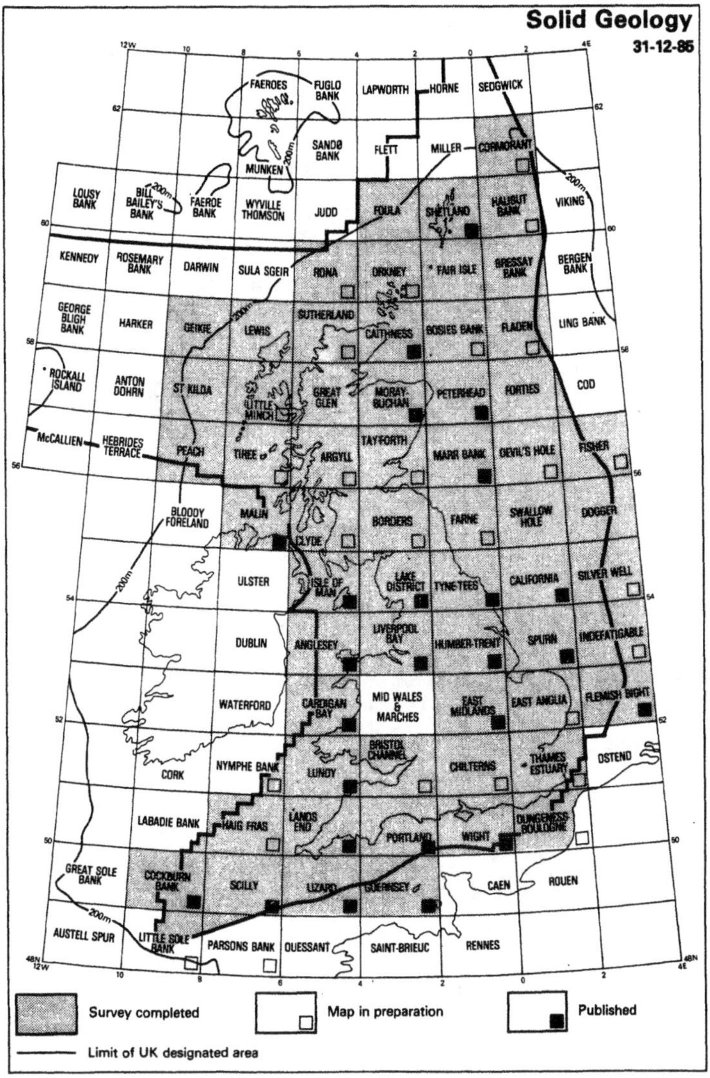

Figure 9. Extent of surveys of solid geology with maps published and in preparation.

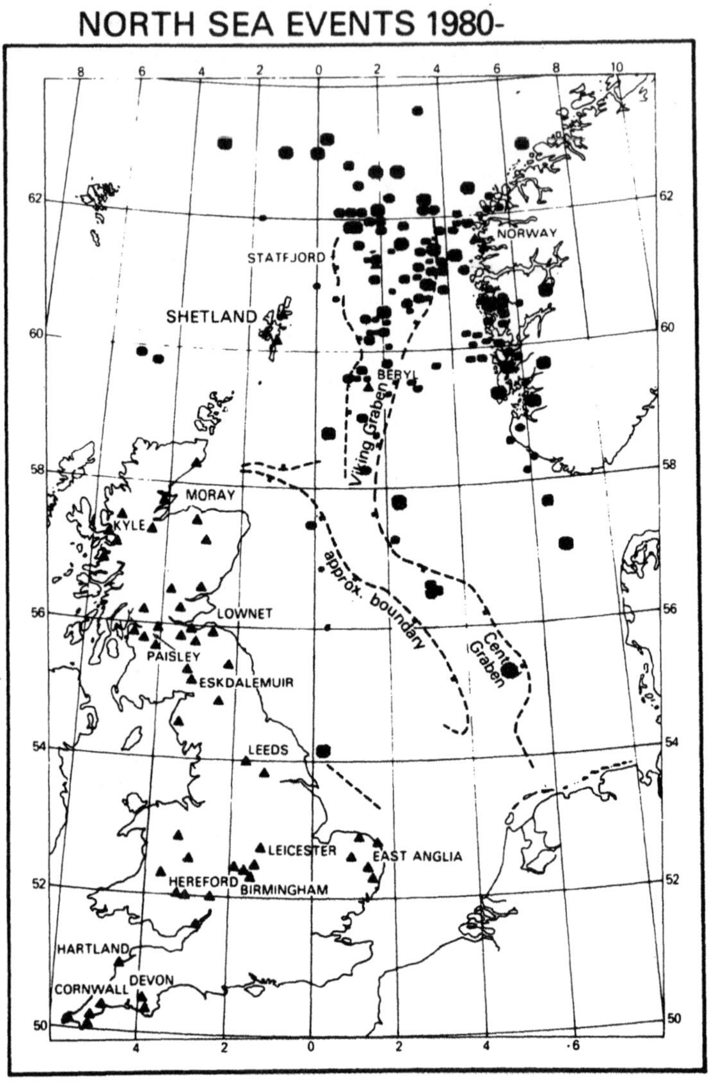

Figure 10. Earthquake epicentres (spots) recorded from the North Sea (1980 to end 1985). The size of the spots indicate the intensity of the earthquakes. The greatest intensity recorded in the period is 4.8 on the Richter scale. (The triangles mark the recording sites).

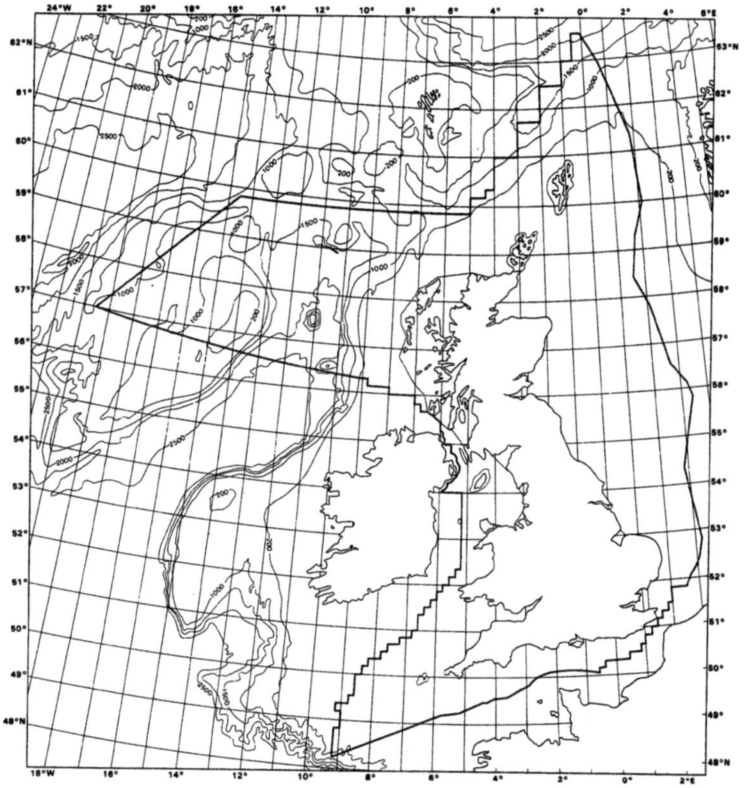

Figure 11. The possible configuration of the EEZ of the United Kingdom.

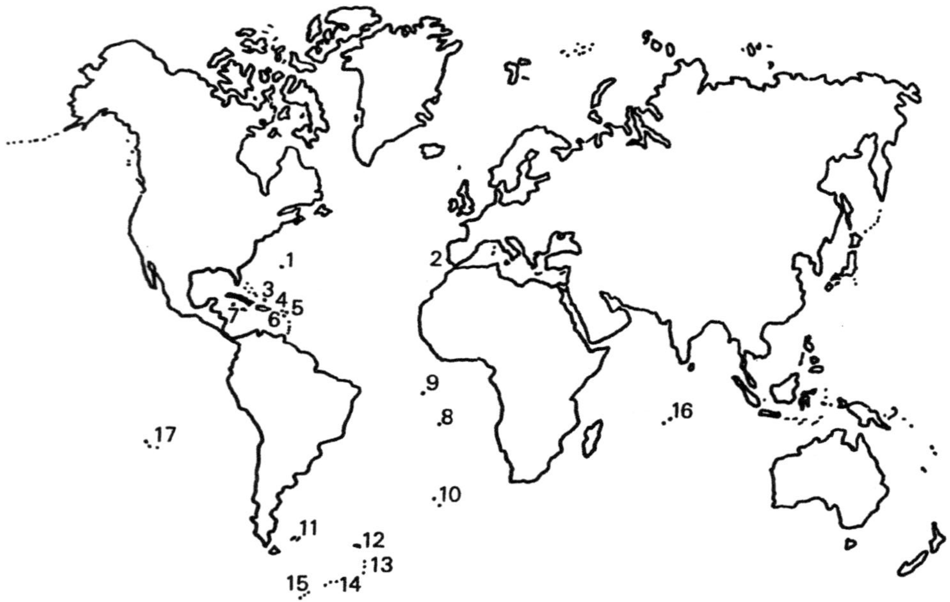

North Atlantic

1. Bermuda
2. Gibraltar

Caribbean

3. Turks and Caicos Islands
4. British Virgin Islands
5. Anguilla
6. Montserrat
7. Cayman Islands

South Atlantic

8. St. Helena
9. Ascension
10. Tristan da Cunha

South-western Atlantic

11. Falkland Islands
12. South Georgia
13. South Sandwich Islands
14. South Orkney Islands
15. South Shetland Islands

Indian Ocean

16. British Indian Ocean Territories

Pacific Ocean

17. Pitcairn Islands

Figure 12. Map of UK Dependencies and Sub-dependencies

THE DEVELOPMENT OF AN INTEGRATED MARINE POLICY FOR THE EEZ : THE CASE OF JAPAN

Professor Michael Gibbons
and
Daniel A. Spagni

The Programme of Policy Research in Engineering,
Science and Technology (PREST)
University of Manchester

The Japanese have been shaped in large part by the land in which they live. Its location, climate and natural resources set the limits and direction of Japanese development. Japan is 50 per cent larger in size than the British Isles, while its population of over 115 million is more than twice our own. Yet geography and geology have not been kind to the Japanese. The whole country is so mountainous that less than a fifth of it is level enough to be suitable as agricultural or industrial land. Japan has by far the highest density of both population and production per square kilometre of habitable land of any country in the world.

Japan is also poorly endowed with natural resources. It is highly dependent on foreign sources of minerals and in particular imports all its nickel, cobalt and manganese. In terms of its primary energy requirements, Japan imports well over three quarters of the coal it consumes, 99% of its oil supply, and has no indigenous uranium production to fuel its growing number of nuclear reactors. Thus, to an almost unparalleled degree amongst industrialised nations, Japan is dependent upon other countries for resources to fuel its prosperity. In order to reduce the impact of such a dependence and promote continued economic and social growth, Japan was one of the first countries to integrate science and technology fully into its basic national economic philosophy.

1. Science and Technology Policy in Japan

1.1 Main Indicators

Japan's expenditure on Science and Technology R&D has expanded rapidly over the past decade with annual increases of between 10 and 15 per cent and is now the third largest in the world after the United States and the Soviet

Union. The total research expenditure (public and private) in Japan for fiscal year 1983/4 amounted to Y6.500 billion (£26 billion). The ratio of R&D expenditure to national income reached 2.8 per cent, again the third largest after the Federal Republic of Germany and the United States.

Industry accounted for 70 per cent of the total R&D expenditure, universities 16 per cent and research institutes 14 per cent. Government funds accounted for 22 per cent and private funds 78 per cent. In 1984 there were 370,000 researchers in Japan, the third largest number after the Soviet Union and the United States. 60 per cent were employed in companies, 30 per cent in universities and 10 per cent in research institutes.

From these few statistics it is already possible to identify some of the main characteristics and originality of Japanese R&D. A fuller picture can be found in a document published every year by the Agency of Industrial Science and Technology and entitled "Trends of Principal Indices on Research and Development Activities in Japan".[1]

1.2 Role of Private Sector

In Japan the private sector plays an extremely important role in determining the level and type of R&D undertaken. Technological innovation has played an increasing role in promoting economic growth in Japan and the dominant view in Industry is that R&D investments bring good dividends with most firms emphasizing priority for R&D (including long term R&D) and development of new technologies and products in their business strategy.

Whereas the growth of research expenditure in the universities and government research institutes in recent years has been rather stagnant (due to the government attempts at reducing the budget deficit), the rate of R&D investment in industry has increased much faster than the growth rate of GNP. Three quarters of the annual R&D budget is funded by private enterprise and the twenty largest firms in Japan account for one third of the nation's industrial R&D spending.

This dominance of the private sector in the funding of R&D has created an imbalance between the level of basic and applied research. The ratio of basic research to the total R&D has declined sharply over the past 10 to 15 years and Japan's total expenditure for basic research is only 30 per cent that of the United States.

An increasing number of firms are turning away from conventional strategies whereby they sought to increase their marketshare by improving their existing products or producing higher value-added products to a strategy that opens up new markets through new technologies and products. In a recent study more than 75 per cent of the firms surveyed were considering entering new fields often representing a radical departure from their traditional markets as a way of strengthening their business and expanding their scale of operation. [2] This often results in a move into leading edge technologies (electronics, life science, new materials) with increased commitment to basic and interdisciplinary research.

The most common form of R&D partnership in Japan is between organisations working in different fields, while cooperation in the United States and Europe usually takes place within the same industrial sector. The flows of R&D funds between industry, university and government in Japan are smaller than in Western Industrialised States and government grants and subsidies to companies in particular are markedly smaller.

However in recent years, Japanese firms have placed a growing emphasis on cooperation with other R&D institutions especially universities and public research institutes. The form of these research agreements is also changing from mere exchange of technical information and technical cooperation to joint or commissioned research.[3]

1.3 Institutional Framework

Japanese Science and Technology policies are formulated and implemented by three government departments: the Science and Technology Agency (STA), the Ministry of International Trade and Industry (MITI) and the Ministy of Education, Science and Culture, the leading role being played by the first two.

MITI is in charge of the development of industrial technologies and is promoting science and technology for industrial applications. MITI's industrial technology policy is carried out by the Agency for Industrial Science and Technology and its Institutes through a variety of large scale research projects classified into 3 categories:

- the R&D programme for the Industrial Bases of the Next Generations which started in 1981 as a 10 year plan;
- the Sunshine project of new energy sources and the Moonlight project of energy conservation systems;
- the National R&D programme usually called the Large Scale Project.

Marine technologies are developed under all three categories. For example Ocean Thermal Energy Conversion and extraction of uranium from seawater are sponsored by the Sunshine Project while manganese nodule mining is part of the Large Scale Project.

The Science and Technology Agency is attached to the Prime Minister's office and is central to the national Science and Technology Policy of the

Japan government. STA has traditionally been seen as a supporter of "big" science (nuclear, space, ocean) but in recent years it has increasingly assumed a position traditionally held by the Ministry of Education as a supporter of small scale basic research. STA is also called upon increasingly to play a more active role in coordinating the Science and Technology policies of the mission-oriented ministries.[4]

In the case of marine Science and Technology, the coordinating role of STA is strengthened by its multiple roles:

- it controls the Japan Marine Science and Technology Centre (JAMSTEC) which is a special public corporation under the direct control of STA. It is financed mainly out of public funds and operates large facilites such as oceanographic ships, submarines and diving chambers.
- it is the secretariat of and it manages the Interministerial Marine Science and Technology Committee which groups all government departments and agencies which are funding marine R&D.
- it is the secretariat of the Council for Ocean Development (COD) which has the task of advising the Prime Minister on all marine matters.

2. Japan Marine Policy : A Review

2.1 General Trends

Because of Japan's lack of land-based resources, the seas have always provided a disproportionate amount of the country's indigenous natural resources and Japan's dependence on the ocean space and its resources is at the highest of any developed State. For example, fish are a vital part of Japanese diet, providing up to 50 per cent of the nation's consumption of animal proteins.

Japan's extensive participation in world wide ocean activities such as fishing and shipping has made it a major beneficiary of the traditional ocean regime dominated by the principle of freedom of the seas.

Until the early seventies, the Japanese government, under pressure from its fishery lobby, maintained a strong opposition to the 200 mile economic zone concept and to other extensive jurisdictional claims made by coastal states. During the debates at the Third United Nations Conference on the Law of the sea, Japan was the only country to oppose actively the emergence of the exclusive economic zone concept and regime; in the event it was not successful.

Failing to achieve its goal of defending the status quo, the Japanese government, recognizing its virtual isolation in global ocean politics, began searching for a new approach which in 1977 resulted in Japan extending its territorial sea to 12 miles and proclaiming a 200 mile exclusive fishery zone. Japan's economic sea area covers 4.5 million sq. km., 12 times its land area and it is the 7th largest in the world.

The process of Japan's adaptation to the modern regime of ocean management took place within the Council for Ocean Development (COD). This consists of an equal proportion of academics, industrialists and government representatives, and has the task of advising the Prime Minister, at his request, on all matters relating to ocean development. In 1978, the Prime Minister submitted the following enquiry to the Council:

> "Japan's land area is small and its resources scarce so that ocean development is expected to make a significant contribution. Recognising the limits placed upon natural resources, many coastal states have set up exclusive economic zones and a new management regime for deep seabed resources. Under these circumstances it is necessary to establish basic policies".5

The Council's response was contained into two major reports, one in 1979 entitled "Basic Concept for Ocean Development from a Long Range Perspective" and in 1980 "Basic Policies for Ocean Development from a Long Range Perspective".[6]

The first report, published in August 1979 forecasted the demands which will be made on the oceans by the year 2000 and identified a series of intermediate goals for 1990. The report was original in that it took a long term rather than short term view of the needs and objectives of ocean development in Japan and that its recommendations were based on a quantitative assessment of the state of the Japanese economy in the year 2000.

The second report, published in 1980, further refined the overall planning exercise and outlined specific tasks which need to be performed in areas which the 1979 Report had identified as requiring government action, ie. development of living resources, seawater, seabed resources, ocean energy and ocean space; coordination of multiple uses of the sea; environmental protection; marine research and survey; development of key technological systems in remote sensing, positioning, new materials, subsea power units; and international collaboration.

In order to rationalise the legal and institutional structure applicable to marine science and technology, the Report suggested that:

- the government should establish an "Ocean Development Commission" in charge of formulating basic policies, coordinating actions and estimating national budget and costs for ocean development.
- The government should enact a "Basic Law for Ocean Development" and other necessary legislations for the management of ocean activities in accordance with the new international ocean regime.

2.2 Japan Marine R&D Budget

Goverment funding (excluding exploitation of oil and gas resources and ship and submersible design) is about £215 million per annum. Private sector funding is over £40 million.[7] By comparison the 1984 Cabinet Office Review gave the total government expenditure on marine science and technology in the UK as £193 million and of this £122.5 million was put down to defence objectives leaving a civil budget of about £70 million (excluding support for university research through UGC).[8]

The size of the budget reflects the fact that Japan is actively promoting a broad range of R&D programmes in the marine field. These programmes are aimed at developing both the technologies necessary in the short/medium term for the exploration of the Japanese economic zone and the technologies necessary in the long term for the exploitation of the many resources lying in the zone.

2.3 Main R&D Programmes

2.3.1 Exploitation of Living Resources

Japan's total fish catch has registered over 10 million tonnes per year since 1975 representing more than 15 per cent of the fish caught throughout the world. With the "200 mile shock" of the late 1970s and tighter enforcement of exclusive fishing zones, Japanese fishing companies have been obliged to operate in an ever decreasing area of open ocean.

Despite national fishing zone extensions, Japan plans to increase its annual fish catch by one to two million tonnes by the beginning of next century in particular for medium and high grade products (eg. lobsters, shrimps, prawns). In order to achieve this objective, Japan's R&D policy in this field is aimed at developing acquaculture and ocean farming including the systematic cultivation of fish, and technologies for efficient breeding and increasing yield of fishery products.

The Ministry of Agriculture, Forestry and Fisheries (MAFF) through its Fishery Agency has been conducting a 9 year study on marine ranching since 1980, together with a systematic plan of developing and rehabilitating coastal fishing grounds. The Agency is trying to establish the technologies needed to produce different marine products, categorised by area, activity range, characteristics and it is also planning to improve the ecological environment created when these products are combined. Trout, mackerel, bluefin tuna, place, scallops and a number of seaweeds have been studied.

The marine and engineering sector is playing an important role with the construction of artificial reefs and breakwaters. The electronic and mechanical engineering industry, the shipbuilding industry and others are also collaborating with MAFF and its research institutes so that fish ranching of salmon, tuna and other high grade migratory species can take place through the establishment of large scale, remotely controlled fishing zones.[9]

2.3.2 Mineral Resources

Extraction of Uranium from Seawater

Extensive laboratory tests in Japan have produced concentrations of uranium close to 3000 ppm - slightly better than that in natural uranium ore. An extraction plant has been built by the Metal Mining Agency of MITI at a cost of £13.2 million and has recently begun operations at Nio-cho (Kagawa Prefecture) on the island of Shikoku. The output is expected to be 10 kg of uranium per year. Although seemingly a small amount, one should note that it took several years for a team at Kiel University in Germany to extract 150 grammes.[10]

Cobalt Crust

The Deep Ocean Minerals Association (DOMA) and JAMSTEC have been conducting for some time research in this field in collaboration with the United States and the Federal Republic of Germany. The fact that some of these deposits could be present within 200 miles of the Japanese coast prompted Prime Minister Nakasone, last September, to establish a committee, under the chairmanship of Dr Okamura, with representatives of industry, government and academia in order to recommend what Japan's future policy in this field should be.[11]

Nodule Mining

The Ministry of International Trade and Industry (MITI) plays the central role in the implementation of Japan's deep seabed resource policy.[12] It has devised a dual system policy which includes:

- a national project-orientated policy aimed at promoting Japan's own resource development;
- a consortium-orientated policy aimed at securing resources through participation in the activities of the international deep sea mining consortia.

In 1973, the Deep Ocean Minerals Association (DOMA) was created at the initiative of MITI to serve as a liaison between the government and the private sector in the development of deep sea resources. Japanese companies have gone into partnership with European and American companies in two international deep sea mining consortia: several Mitsubishi companies formed the Kennecot consortium in 1974, and in June 1975 Sumitomo, together with 22 other companies, established Deep Ocean Mining Co. Ltd. (DOMCO) and finally joined the the Inco consortium. In December 1980, MITI and the Ministry of Finance agreed to adopt manganese nodule mining technology as a Large Scale Project, with a budget of £80 million over 9 years and the establishment of a partnership organisation to administer the R&D work. All the companies active in this field have now joined this organisation and in 1982 established the Laboratory of Managanese Nodule System at the National Research Institute for Pollution and Resources in Tsukuba. The laboratory has a 200 m deep vertical water tank to carry out basic hydraulic and air lift tests.[13]

Hydrothermal Deposits

In May 1984, the Resource Council, an advisory organ of the Director General of the Science and Technology Agency (STA) submitted a report recommending the survey and study of hydrochemical deposits in the southern waters of Japan. This was followed by a series of surveys conducted by STA through JAMSTEC, and by MITI through its Metal Mining Agency. Japan is also actively collaborating in this field with France (exploration of the Japanese Trench by the Kaiko programme) and the United States (Pacific Ocean Exploration).

2.3.3 Ocean Energies

Ocean Thermal Energy Conversion (OTEC)

Research and Development on OTEC in Japan are conducted by the government and by private industry. The government programme is part of the MITI Sunshine Project aimed at developing domestic energy resources. It is based on the Electro-technical Laboratory of the Agency of Industrial Science and Technology and to a lesser extent on Saga University in Kyushu.[14] A number of companies are also involved, including Toshiba. Based on extensive testing of the small laboratory facility of the Electro-technical Laboratory at Tsukuba the aim of the project is to build a 1 to 3 mw pilot plant.

Most of the R&D activity on OTEC in Japan has been conducted by private groups. These include the experiment on a 100 kw test plant in Nauru conducted by the Tokyo Electric Power Services Company in 1981 and the smaller experiment conducted by the Kyushu Electric Power Company in Southern Japan.

Wave Power

Apart from OTEC, wave energy is the most attractive of the ocean energy options being pursued in Japan. JAMSTEC has been conducting R&D into wave power since its creation in 1971. Most of the work has been concentrated on the Kaimei system. Since 1979, the Kaimei experiments and sea trials have been carried out as a joint international project of the International Energy Agency (IEA).[15]

A small coastal wave power generation system was operated successfully in 1983. The plant was developed by the Research Development Corporation of Japan and the test was conducted by Fuji Electronic Company Ltd. and Mitsui Engineering and Shipbuilding Company Ltd. in collaboration with JAMSTEC. The Institute for Technical Research of Ports and Harbours of the Ministry of Transport has also investigated the incorporation of wave power plants into breakwater designs.

Ocean Currents

The Science and Technology Agency has funded for the past 10 years a research programme on the study and the exploitation of the Kuroshio warm ocean current. Conceptual work has been carried out at the National Institute of Resources on an ocean current power generator system to be used in the Kuroshio flow. Japan has recently set up with China a 7 year collaboration project of study of the East China Sea and the Kuroshio current.

2.3.4 Ocean Space : Artificial Islands and Coastal Developments

As was mentioned earlier, Japan is extremely crowded and becoming more so. Land is extremely expensive and coastal areas, representing 30 per cent of the total habitable space, account for 50 per cent of the population and 60 per cent of industrial production. Offshore "space" is therefore seen as an enormous potential resource.

The small coal mining islands in the 1950s were rapidly followed by spectacular projects such as Port Island off Kobe which has now 20,000 permanent residents and is the world's largest container port, and the construction of scores of islands in Tokyo Bay which have enabled many companies to relocate and rationalise their plants (eg. Ogishima). The island airports of Nagasaki and Haneda are to be followed this year by the start of the construction of the New Kansai International Airport. The momentum for new artificial island projects is coming from MITI, the Ministry of Transport, Keindanran (the Federation of Economic Organisations) and in particular its Kozai club which groups the major steel companies.

Most of Japan's major shipbuilders are also active in the construction of floating industrial plants in particular for developing countries. These include desalination plants, floating power plants and floating pulp plants.

2.3.5 Ocean Survey and Studies

In order to increase its capability of exploring its home waters and the deep seabed, Japan has pursued a vigorous policy of investment in new and advanced oceanographic facilities and of promoting the development of a whole range of oceanographic instruments.

Unmanned Submersibles

The development programme for remotely operated vehicles (ROV) is conducted by JAMSTEC and the vehicles developed up to now include the JTV vehicles which can operate down to 200 m and the HORNET which can operate down to 500 m. The DOLPHIN 3 K with an operating depth of 3,300 m is due to be completed before the end of the year.

Manned Submersibles

The Council for Ocean Development suggested in 1969 that Japan should develop a 6000 m class submersible. Research began in 1976 on a slightly less ambitious 2000 m depth submersible. The deep sea submersible SHINKAI 2000 and its support vessel NATSUSHIMA were both completed in 1981 and delivered to JAMSTEC. Work on a submersible with a deeper diving capability was started by STA in 1985 and is due to be completed by 1989. With a maximum operating depth of 6,500 m this submersible will compete with the US Seacliff and the French Nautile. Preliminary studies were completed in 1984 and work on the basic design started in 1985.

Marine Remote Sensing

The Japanese ocean remote sensing programme is based on the experimental MOS 1 Marisat spacecraft which is due to be launched later this year. The satellite will be placed into sun-synchronous orbit and equipped with three new observation systems (all made in Japan) for detecting changes in the sea surfact temperature, changes in the colour of the ocean and measurement of water vapour in the atmosphere.

Ocean Survey Vessel

In May 1985 JAMSTEC took charge of the oceanographic, ocean survey vessel KAIYO. This is a semi-submerged catamaran vessel which provides a large deck area and is equipped with multi narrow beam sonar (Seabeam) and a new Japanese-designed dynamic positioning system. KAIYO is used as the main support for Japan's diving programme (New SEATOPIA Project).

Oceanographic Instrumentation

Since 1983 the STA has been funding a programme entitled "Basic Technology for Oceanographic Instrumentation Research" aimed at establishing a system for effective marine surveys in Japan's economic zone. Work is carried out both by private companies and by the government. The systems which are currently being studied include optical fibre sensors, acoustic tomography, buoy systems for oceanic observation and fisheries mapping, deep sea topographical data analyser for Seabeam and a robot capable of monitoring underwater volcanoes.[16]

2.3.6 International Programmes

Japan is involved in 3 major bilateral collaboration programmes, with the United States, France and the Federal Republic of Germany, involving the exchange of scientific, technical information and researchers and joint scientific projects.[17]

These bilateral agreements have enabled Japan to conduct research on deep sea hydrothermal vents and associated sulphide deposits without a national deep sea manned submersible capability but using the US Alvey and French Naulite (Kaiko / exploration of the Japan Trench) collaboration with China has recently been stepped up with a 7 year study of the East China Sea and in particular the Kuroshio ocean current system.

Multinational cooperation programmes include the West Pacific Ocean Joint Research (WESTPAC) and the Ocean Drilling Project. WESTPAC include the Kaimei wave experiment at JAMSTEC which is assisted by the International Energy Agency of the OECD.

3. Conclusion

There is broad agreement in Japan that the country should be geared up to develop the knowledge and technology necessary to exploit its own 200 mile economic zone, once established.

Japan is still in the process of formulating an integrated policy towards its economic zone. An ambitious marine Science and Technology programme is strongly supported by both government and private industry.

This programme benefits from the strong institutional structure which already exists within the government for the development and implementation of its Science and Technology Policy. The mechanisms of coordination and information transfer are already there and the government's action suggests it feels the individual ministries and agencies can deal with the present problems adequately.

The following main features are already apparent:

- government departments and agencies play a major role;
- the development of surveying procedures and technologies is seen as the first step towards the assessment of the potential of the zone;
- the private sector is extremely supportive of most programmes;
- the ultimate aim of the government marine S & T policy is the formulation of an integrated, managed policy which allows for the coexistence of the multiple demands on the region and its optimal development.

References

1. Trends of Principal Indices on Research and Development Activities in Japan. Revised Edition. Agency of Industrial Science and Technology. February 1985.

2. Science and Technology White Paper. New R&D Development in the Era of Cooperation. Science and Technology in Japan. Vol.5, No.18. April-June 1986, pp.7-13.

3. Towards the Creation of New Technology for the Twenty-first Century. White Paper on Science and Technology. Science and Technology in Japan. No.14. April-June 1985, pp.25-31.

4. Kodama, F. Science and Technology Policy Outlook: Main Trends and Issues - Japanese Case Study. Institute for Policy Science Research Report, Saitama University, 1980.

5. Nakahara, H. Consensus building in the Council for Ocean Development, in: Friedheim, R. et al., Japan and the New Ocean Regime, Westview Press, 1984 at p.94.

6. A Summary of the Reports of the Council for Ocean Development, 1978-1980, Science and Technology Agency - mimeo.

7. House of Lords, Session 1985-86. Select Committee on Science and Technology - Marine Science Technology, Vol.III, p.96, HMSO, London.

8. Cabinet Office - Annual Review of Governmental Funded R&D, 1984, HMSO, London.

9. Ford, G. et al. Japan and New Marine Technology, Marine Technology, Brief No.4, PREST, University of Manchester, December 1983.

10. Nature, vol.320, 10 April 1986, p.475.

11. Personal Communication.

12. For an in-depth review of Japan's deep sea mining policy see: Georghiou, L., et al. Japanese New Technology - Creating Future Marine Industries, Marine Policy, October 1983, pp.239-253.

13. Technocrat, vol. 17, no.4, April 1984.

14. Japan's Sunshine Project - Summary of Comprehensive Research, Agency of Industrial Science and Technology (MITI), 1985.

15. Ishii, S. The JAMSTEC Wave Power Generation Projects; Japanese - French symposium on Ocean Developments, Tokyo, 9 - 10 September 1985 (Japanese French Committee on Cooperation on Ocean Developments).

16. Ocean Development in Japan. Japan Industrial and Technological Bulletin, Special Issue, no.20, 1984, (Japan External Trade Organisation).

17. International Cooperation on Science and Technology Science and Technology in Japan, vol. 5, no.18, April - June 1986, pp.14-15.

THE MANAGEMENT OF AN EXCLUSIVE ECONOMIC ZONE (EEZ): THE ROLE OF THE MARINE RESOURCE ATLAS

J W Ramster
Ministry of Agriculture, Fisheries and Food, Directorate of Fisheries Research, Fisheries Laboratory, LOWESTOFT, Suffolk, NR33 OHT

INTRODUCTION

Marine Resource Atlases are best described as organised collections of maps of all aspects of specific sea or oceanic areas. Their breadth of subject matter and broad-brush approach, which are sometimes scorned by "experts", are positive advantages because they allow the user whether it be a member of the public, a government official, a teacher, a politician, a scientist or an industrialist, to grasp easily and conveniently the elements that go to make up the region under review. They are, in short, the essential introduction to someone taking an interest in an area for the first time as well as a subsequent look-up facility. All without exception provide guidance as to where more detailed maps or studies can be found. Some have directly associated with them books which not only describe the mapped features but also discuss the relationships between them, e.g. Miles and Fluharty, 1984: most have notes alongside each map, e.g. Lee and Ramster, 1981 or as an introductory section, e.g. Caddy and Garcia, in press. In varying degrees therefore Marine Resource Atlases are analytical tools that can help in the evolution of regional management policies.

It was not always so. Indeed, until about 1975 Marine Resource Atlases did not exist. There were marine atlases that concentrated on some aspect of the physical or biological or commercial worlds but none that focussed on a particular marine area and aimed to put between two covers in map form all that there was to know about it. The stimulus in the United Kingdom to providing such a book was given by the experiences of Arthur Lee, the Deputy Director of Fisheries Research for England and Wales in the early 1970's, as he was called to attend the various committees grappling with the need to set up "systems" that would manage marine aggregate extraction and the then new oil and gas fields in the UK sector of the North Sea. Time after time he found these committees discussing aspects of the problems which they faced without the use of the maps that were "common coin", as it were, to physical oceanographers and fisheries biologists. Consequently, a MAFF "Atlas of the Seas Around the British Isles" was produced in a black-and-white, A4 format in 1975 and distributed free of charge to interested parties. Extra pages were added in 1977 and 1979. At the last count some 1000 copies were in circulation with most of these distributed amongst oil-related companies, government departments and schools. Since 1975, there has been a succession of these atlases for areas as diverse as Narragansett Bay, Rhode Island, the Eastern USA coastal and ocean zone, the seas around New Zealand, the North Pacific, South East Asian seas, Eastern Canada, the Gulf of Mexico, the Danish coastal zone, the seas off Finland and a coloured A3 update of the original MAFF Atlas. (See Annex 1). Three thousand copies of the latter book are now in circulation.

Much of the interest in these atlases stems from the evolution into case law and the status quo of the concept of the Exclusive Economic Zone (EEZ) (Annex 2). Each of the 139 coastal states - 103 had formally declared EEZ's at December 1985 - now needs to know what resources are available in the water and on and under the sea bed of its EEZ if it is to develop and manage them properly. Maps remain the most convenient and assimilable means of summarising the results of the expensive research vessel cruises needed to determine what is or is not present and organised collections of such maps, i.e. atlases, follow. Knowledge being the amorphous matter that it is, the existence of maps, such as for example a circulation pattern, calls for a complementary presentation of the distribution of current meter moorings that made up the database from which it was drawn. Less esoterically a map of average surface temperatures by season complements charts of fish distributions. Hence, a Marine Resource Atlas takes shape as rather more than just a compilation of the results of a series of cruises investigating the "Inner Space" of an EEZ. Another commonly found, unexpected result of the initial effort to produce such an atlas, is the sudden realisation of just what aspects are still virtually unknown, or are represented in a bogusly unambiguous way. Atlas-makers have found that it is worthwhile mapping a feature well-known in one part of a region, even though similar quality data for another part is simply not available, because of the focus that it gives to what is needed for a better mapping exercise in the future. Such maps are always circulated with a warning note about the variable data quality.

The fact that there is a large degree of commonality of experience amongst Marine Resource Atlas-producers came to light most clearly in October 1985 when a group of them met in London for two days under the auspices of the Royal Geographical Society and the Challenger Society (Annex

3). Time and again the point was made that the usefulness of the atlas format resulted from a balance being struck between scale, colour, the ratio of text to maps the physical size of the book being produced and the resources available for production. The other dominant theme to emerge was the need to meet head-on the fact that some aspects of any atlas are out of date before the book is published. Editors can minimize the importance of this fact of life to some extent by carefully choosing the data presented (e.g. 10-year trends of fish catches rather than values for one particular year) or the topics covered (e.g. the geological structures underlying offshore oil fields as well as the pattern of rigs and platforms at a particular time). However, some users may well need the current distributions of this or that factor if they are to gain maximum use from an atlas project.

This calls, of course, for computerised databases which have been seen from the first (see Lee and Ramster, 1981) as the essential foundation of the fully developed Marine Resource Atlas. Fortunately, the Ocean Assessments Division (OAD) of the US National Oceanic and Atmospheric Administration (NOAA), the US Geological Survey and the British Geological Survey (BGS) respectively were able to demonstrate the practicability of such systems at the meeting. The OAD's programme of work is the most broadly based of the three examples cited. Indeed, on the evidence presented of the whole series of US Marine Resource Atlases that are in production, it is a text-book example to all other countries of how things should be done in kind if not degree. It was therefore particularly important to learn that, in the OAD's experience, digital mapping techniques should be viewed as just another set of tools to be used by the atlas maker rather than ends in themselves. To workers who are trying to get into this field such warnings are, of course, of slight importance compared with the likely gains that will stem from the ability to manipulate and plot a large data-set in various ways before deciding which best meets the need of the moment, and then being able to update this exercise in real time. For them the most important point was that the OAD software is deemed to be non-proprietary in principle and could be demonstrated and exported to other interested parties. It is of considerable potential importance, therefore, to be clear as to just what has been achieved in some 7 years.

The Ocean Assessments Division of NOAA was set up in 1979 and has evolved until today it consists of about 140 people with an annual budget of

$20 million: some 20 people and $2 million a year spread over several of its sections are given over to Marine Resource Atlas production. The reason why such resources have been deployed is that it is felt that better decisions will be made about how to use the EEZ in general, and the coastal zone in particular, if the information on which they are based is sound and up-to-date (Ehler and Basta, 1984). Consequently, the OAD has been organising and making readily accessible to all the best available information on the characteristics of the US EEZ. As from 1984, for example, there has been a systematic national programme conducted by NOAA and managed by the OAD, to produce good data via private contracts, as well as ships and laboratories belonging to NOAA, that relate specifically to the environmental health of the region. Are conditions getting better or worse? How does one estuary compare with another? How does the coastal zone compare with the continental edge? The investment is undoubtedly substantial but it is small relative to the productivity of the natural resources being monitored and the potential costs of clean-up operations if a long-term problem is not picked up at an early stage. From the first, fortunately, OAD realised that they were moving from a period when there was little or no data relating to the EEZ to a time when there could be too much to deal with if computerised databases and cartographic techniques were not employed. As a result, a system has been evolved that can be virtually point specific but equally can be addressed in terms of general trends in time or space. This means that for the enquirer a whole range of outputs is available for viewing prior to hard-copy production or after a broad-brush atlas map has been perused.

TANGIBLE ACHIEVEMENTS TO DATE

Who is asking questions of the system and what kind of questions are being posed? The enquirers are, in the main, Departments of the Federal Government who are contemplating or interpreting legislation, the staff of Senators and Congressmen who are pushing projects of various kinds, private lobbyists acting for "interests" of all kinds and industry itself considering future investments. One recent query concerned the possibility of changing the rules in relation to the characteristics of the "produced-waters" pumped overboard from oil rigs positioned in the US EEZ. To give some idea of the scale of investment involved there were 4056 rigs of various kinds working in the Gulf of Mexico section in 1983. A series of tests showed the list of marine animals and plants that would be harmed by a range of differently constituted "produced-waters". A search of the database showed those areas of the shelf where there was significantly less risk of items on this list being present in this or that type of "produced-water". Now there are different rules for different areas with an estimated saving to the industry of some $500 million per year. Professionals had performed the long-term background monitoring test phase, entries to the database had been scrutinised over time by both OAD professionals and external peer review and within a week of the query being made, a view was formed that influenced policy making decisively.

A rather more jaundiced view of the Atlas of the North Pacific was taken by West Coast oil and gas executives when they were given a presentation that ended with a request for financial support from the non-OAD editors in this case. "Why should we support a project that will probably lead to extra trouble and expense for us in due course?" was the line taken (D Fluharty, pers. comm.). It would appear that the extra problems which they were shying away from here are to stay. In October 1985 the point was made by one speaker that a recent OECD Opinion Poll showed that some 60% of both Americans and West Europeans rated environmental protection higher in importance than economic growth. Dr Martin Holdgate, Chief Scientist at the Department of the Environment, had already used this result as an argument for greater informed public debate on pollution matters. "The community wants to be able to check for itself the validity of the decisions taken on its behalf". "An effort to communicate to the larger public is essential if intelligent people are going to get full access to the facts in a fashion impossible in short newspaper articles" (The Times, 12 September 1985).

The West Coast oilmen who shrugged off their friendly, neighbourhood, Marine Resource Atlas would do well to look at the history of the Lancaster Sound Project in the High Arctic of Canada (Dobbin _et al._, 1983). Here, a public inquiry sparked off by proposed exploratory drilling recommended that an objective assessment be made of the best possible use of the area to the year 2000 given the potential conflicts between the teeming marine life, new year - round shipping through the North West Passage, proposed oil, gas and metals exploration, increased tourism and the centuries - old life-style of the Inuit Indians which depends on the natural environment and is in harmony with it. Underpinning all the planning reports was a Marine Resource Atlas by James Dobbin Associates having 83 maps and texts in English, French and Inuktitut. Behind this was a computerised database that allowed composite maps to be produced on request. The whole exercise - including a period of review via public meetings - took some 4 years and ended with the Northern Land Use Planning Agreement (Canada, Dept. of Indian and Northern Affairs, 198 which takes full account of the intricate relationships between the coastal an marine zones.

The same firm has used a similar assemblage of techniques for resiting a small population of endangered sea otters on the Californian coast for the US Department of the Interior. For the Canadian Government, on the other hand, it developed a Gulf of Maine Atlas that was used throughout the Oral Pleadings before the International Court of Justice as the USA and Canada cleared up a long-standing maritime boundary dispute. It was also submitted as evidence in two of the Written Pleas. James Dobbin argues that his maps were a significant factor in the final judgement that left 60% of the scallop grounds on the George's Bank - a fishery worth $60 million a year - in Canadian hands. This may well be a factor too in Canadian plans to set up national Marine Resource Databases, and an equivalent of the OAD, at a cost of about $0.4 million. On a smaller scale in a database sense perhaps, but nevertheless with world-wide applications, FAO is now seeking to spread the gospel of resource databases and linked mapping throughout the developing countries (Caddy and Garcia, in press). It is looking for micro-based software that will allow rapid updating and production of maps that are easy to interpret. Above all, FAO recognises from experience that it is no good waiting for blanks in a database to be filled before getting started on the maps: such blanks as exist have to be seen as a spur to future action.

In much the same way, the management within the Ocean Assessments Division of NOAA has grasped the point that it must not wait for people to come and ask questions of a passive system: by producing Marine Resource Atlases that organise, compile and synthesise the details it can communicate most effectively large amounts of complex technical information and indicate to interested parties that more detail is available if required. In the time that MAFF has sat on its laurels in regard to its "Atlas of the Seas Around the British Isles," OAD has produced new atlases of the East Coast, the Gulf of Mexico, the Bering Strait area, the West Coast of the USA, the

Gulf of Alaska and, in even broader brush, a national EEZ resource atlas that is derived from the other products. It has also produced an estuarine atlas that takes the 82 major estuaries in turn and describes their physical and chemical characteristics and a biological equivalent is on the stocks. Such productivity is possible because the system and techniques evolved for one area - the East coast of the USA - are just as applicable and relevant to all other areas. The atlases are the single most visible product of the programme but their underpinning of computerised databases and analytical and cartographic techniques is even more important. (Annex 4).

THE SITUATION IN THE UK

If the UK is to manage its potential EEZ effectively it seems clear that we need to develop a similar capability to that of the OAD. In fact we are not all that far away from doing so in theory. The British Geological Survey already has several years of experience in this field related to its systematic exploration of the UK continental shelf areas. The Natural Environment Research Council's (NERC's) Marine Information Advisory Service (MIAS) has acted as a national data centre for the physical and chemical sciences for a decade now and is highly computerised. Moreover NERC, MAFF and DAFS each have large datasets relating to marine biological phenomena as computer archives. It may be just a question of putting our "marine data resources"-house in order, an echo perhaps of the fragmentation in marine matters in general found by the House of Lords that reported on Marine Science and Technology in 1986.

A pivotal role in this exercise, if it ever got under way, might be played by NERC's Unit of Thematic Information Systems at Reading where specific research aims include:

i) development of methods of spatial data management, for improved storage, retrieval and manipulation;

ii) development of automated geographic information systems involving the development of techniques for the integration of the various forms of environmental data;

iii) development of improved methods of cartographically displaying data; and

iv) rigorous evaluation of statistical and mathematical techniques, used in the analysis of the data.

Moreover, 40% of its work is to be concerned with collaborative projects specifically selected to explore problems which have wide application. The creation of a national Marine Resource Database and consequent map production at all scales, culminating perhaps in a series of atlases, would seem to be an area of work that this unit might address with help from MIAS, BGS, MAFF and DAFS and all other relevant bodies in the UK, where appropriate.

In the same way as MAFF found in the 1970's, that it was just as easy to map the seas around Britain as it was to consider the North Sea alone, the proposed UK-equivalent of the OAD could find itself dealing with this topic in a European Community (EC) rather than a national context. This may seem to be too large a step to take from the first but it should not be left out of any initial planning. Through ICES and the Oslo and Paris Commissions the rudiments of the required structure for data pooling already exist to be picked up by the members of the EC. The problems may seem large but are they any greater than those faced by the US in 1979 if the political will to fund and sustain the project exists? With astute lobbying, in fact, the necessary funds could be provided in large measure by the Community, so raising the UK's profile there as receptors of research funding and thus again following one of the recommendations of the House of Lord's Report. (House of Lords, 1986).

SEA-USE MAPS: A POSSIBLE SPIN-OFF FROM MARINE RESOURCE DATABASES

It has been emphasised several times that once a Marine Resource Database has been created the individual user at a terminal can, in the first instance, determine what ephemeral products he wishes to produce and at what scales. This facility was looked at from two separate viewpoints in October 1985. One speaker emphasised the freedom provided to investigate statistically as well as spatially the possible relationships between various factors. Another suggested that the existence of these systems highlighted the hollowness of that approach and forced us to evolve models of marine processes of all kinds that would be tested by the data that was available (Carlton Ray, 1985). To some extent he was echoing the OAD view that the computerised facilities are only a tool and not an end in themselves. However, there is no denying that the ability to organise and manipulate data easily is a very attractive proposition and one particular area in which it could be practically useful very quickly has surfaced in the discussions of the Standing Marine Resources Committee of the Royal

Institution of Chartered Surveyors (RICS). There, the members have noted that their basic land-management tool, an Ordnance Survey Land Use map drawn to the appropriate scale for their immediate needs, does not have a marine equivalent. They wonder if, therefore, Sea-Use maps of particular areas may be marketable. Any one sheet would draw together those features of marine resources that were of particular importance to potential developers. Thus, while such things as tidal streams might be dealt with by reference to a tidal stream atlas the distribution of herring eggs and potential gravel dredging areas, for example, would be shown to highlight a potential "conflict of use". This idea was looked at too in some detail by Arthur Lee in the mid-1970's in an ICES context but did not gain support from other scientists (Lee, 1976). A sub-group of the RICS committee is now looking at it again. It cannot be stressed too highly that the judgement of what is and what is not important in such cases stems very much from the standpoint of the individual map-maker and so the great virtuosity of a computerised system comes into its own again, by making all such standpoints easily accessible.

CONCLUSIONS

This survey of the role of the Marine Resource Atlas in helping to manage an EEZ has necessarily been wide-ranging but the main points made are relatively few in number. First, Marine Resource Atlases have become popular because coastal states need easily accessible compendiums of information about their EEZ's. Second, the mass of information available already, and the need to keep it updated for policy makers, necessitates the use of computerised databases and cartographic techniques. Third, there are no insurmountable technical problems as far as these facilities are concerned: various groups, and most particularly the OAD of NOAA, have blazed the trail. Finally, the UK should bring together various strands in its existing marine sciences structure that would allow a similar programme to that followed by the USA to flower and should seek, at the same time, to play a leading role in the EC in such matters.

Annex 1. Marine Resource Atlases displayed at the Marine Atlas Colloquium, London, October 1985

Atlas of Finland

Atlas of Fisheries of West and Central Mediterranean

Atlas of Living Recources of the Seas

Atlas of Major Atlantic Coastal Fisheries and Invertebrate Resources Adjacent to the Canada/US Boundary Area

Atlas for Marine Policy in SE Asian Seas

Atlas of Marine use in the North Pacific Regions

Atlas of the Seas Around the British Isles

Canadian Atlantic Offshore Fishery Atlas

Channel Islands - National Marine Sanctuary - Data Atlas

Coast of Scotland

Directorate of Fisheries Research Technical Report No 20

Eastern US Coastal and Ocean Zones

General Bathymetric Chart of the Oceans - GEBCO, 5th Edition

Geological-Geophysical Atlas of the Indian Ocean

Gulf of Mexico Coastal and Ocean Zones

Gulf of St Lawrence - Water Uses and Related Activities

Interpretive Atlas of Narragansett Bay

Lancaster Sound Region - Data Atlas

Lancaster Sound Region, 1980-2000

Mitchell Beazley Atlas of the Oceans

New Zealand Atlas of Coastal Resources

Nova Scotia Fisheries Atlas

Passamaquoddy Marine Research Pilot Study - 1982 Fisheries Data Atlas

Piscatorial Atlas - OT Olsen

Planning & Marine Conservation Strategy for the Caribbean Region

Rand McNally Atlas of the Oceans

Sensitivity Maps for the Canadian Atlantic Fishery

Serial Atlas of the Marine Environment

Times Atlas of the Oceans

Note: For further details consult your Librarian or The Librarian, MAFF, Fisheries Laboratory, Lowestoft, Suffolk, NR33 OHT, Tel: 0502-62244,
Tlx: 97470
FAX: UK 0502-513865

Annex 2. Changes in attitude towards EEZ's

ELIZABETH I TO THE SPANISH AMBASSADOR, 1580

"The use of the sea and air is common to all: neither can a title to the ocean belong to any people or private persons, forasmuch as neither nature nor public use and custom permit any possession thereof".

PRESIDENT REAGAN TO THE WORLD, 1983

PROCLAMATION 5030 OF MARCH 10, 1983
EXCLUSIVE ECONOMIC ZONE OF THE UNITED STATES OF AMERICA
BY THE PRESIDENT OF THE UNITED STATES OF AMERICA

A PROCLAMATION

"whereas the Government of the United States of America desires to facilitate the wise development and use of the oceans consistent with international law;

whereas international law recognizes that, in a zone beyond its territory and adjacent to its territorial sea, known as the Exclusive Economic Zone, a coastal state may assert certain sovereign rights over natural resources and related jurisdiction; and

whereas the establishment of an Exclusive Economic Zone by the United States will advance the development of ocean resources and promote the protection of the marine environment, while not affecting other lawful uses of the zone, including the freedoms of navigation and overflight, by other states;

now, therefore, I, Ronald Reagan, by the authority vested in me as President by the Constitution and Laws of the United States of America, do hereby proclaim the sovereign rights and jurisdiction of the United States of America and confirm also the rights and freedoms of all states with an Exclusive Economic Zone, as described herein . . . .

. . . . in witness whereof, I have hereunto set my hand this tenth day of March, in the Year of Our Lord Nineteen Hundred and Eighty-three, and of the Independence of the United States of America the Two Hundred and Seventh."

Annex 3. Marine Resource Atlases, 3-4 October 1985: List of speakers and presentations given

J Caddy (Marine Resources Service, Fisheries Resources and Environment Division, FAO). Mapping as a Tool for Fishery Management at Regional or National Level.

G Carleton Ray (Research Professor, Department of Environmental Sciences, University of Virginia). The Analysis of Mapped Data: Some Possibilities and Pitfalls.

J A Dobbin (President, James Dobbin Associates Inc, Coastal and Ocean Planners). The Development of Marine Data Atlases for International Maritime Boundary Delimitations: the Gulf of Maine Case.

C N Ehler (Chief, Ocean Assessment Division (OAD), NOAA) and D J Basta (Chief, Strategic Assessment Branch OAD, NOAA). The Role of Marine Data Atlases in the Strategic Assessment of Ocean-use Conflicts.

E Escowitz (US Geological Survey). Mapping the Resources of the EEZ of the United States.

C Evans (British Geological Survey). The British Geological Survey's Data Acquisition and Map Production Programmes Relating to the Continental Shelf of the United Kingdom.

Susan Gubbay (Marine Conservation Society, UK). The Coastal Directory of the United Kingdom: a Way Ahead for Marine Conservation.

A J Lee (former Director of Fisheries Research for England and Wales). Marine Resource Atlases: the Conceptual Framework with Special Reference to the Atlas of the Seas around the British Isles.

E Miles and D Fluharty (Institute for Marine Studies, University of Washington, Seattle). The North Pacific Atlas: its Origin and Future use.

D J Scarratt (Director, St Andrews Biological Station, New Brunswick, Canada) and A Kerr (Canadian Hydrographic Service, BIO). Marine Resource Atlases in Canada.

D P D Scott (Leader, UK Delegation to the IOC). Ocean Mapping.

M J Valencia (Research Associate, East-West Resource Systems Institute, Hawaii). The Atlas for Marine Policy in Southwest Asian Seas: an Overview.

C Wooldridge (Department of Maritime Studies, UWIST). The Times' Atlas of the Oceans: its Objectives and Market.

Annex 4. OAD databases 1986

Coastal Outdoor Recreation: estimates economic value of the recreation services provided by coastal and oceanic areas.

Coastal Wetlands: extent of salt marshes, fresh marshes, tidal sewages and flats for contiguous USA.

Estuaries: summaries of the physical and chemical characteristics of 82 major estuaries.

Marine Living Resources: life histories of 300 species stored.

National Coastal Pollutant Discharge Inventory: characteristics of present sites and prohibited areas listed for marine counties and open seas. Contains details of 23 commercial fish and 15 invertebrates.

National Status and Trends: 150 estuaries and coastal sites monitored for 17 elements, 33 aromatic hydrocarbons and 6 PCB's.

Shellfish Register of Classified Estuarine Waters: links shellfish communities with their habitats.

REFERENCES

CADDY, J.F. and GARCIA, S. (In press). Fisheries Thematic Mapping: a Prerequisite for Intelligent Management and Development of Fisheries. FAO Fish. Tech. Pap.

CANADA, DEPARTMENT OF INDIAN and NORTHERN AFFAIRS (1983). Northern Land Use Planning in the North-West Territory: Basis of Agreement Between the Department of Indian and Northern Affairs and the Government of the North-West Territory, July 28 1983. Department of Indian and Northern Affairs, Ottawa, 8 pp.

CARLTON RAY, R.G. (1985). Man and the Sea: the Ecological Challenge. Am. Zool., 25: 451-468.

DOBBIN, J.A., LEMAY, M.H. and DOBBIN, N.E. (1983). Lancaster Sound Regional Study: Coastal and Marine Resource Planning in the Canadian High Arctic. J. Cstl Zone Manag., 11 (1-2) 71-89.

EHLER, C.N. and BASTA, D.J. (1984). Strategic Assessment of Multiple Resource - use Conflicts in the US Exclusive Economic Zone. Proc. Oceans '84, 1-6.

HOUSE OF LORDS (1986). Marine Science and Technology. Vol 1. Report. HMSO, London, 96 pp.

LEE, A.J., DICKSON, R.R. and MILLNER, R.S. (1977). ICES Working Group on the Effects on Fisheries of Marine Sand and Gravel Extraction: Report of Co-ordinators of Material Supplied in Response to Council Resolution 1976/H:14. ICES CM 1977/E:68, 11 pp (mimeo).

LEE, A.J. and RAMSTER, J.W. (Compilers) (1981). Atlas of the Seas Around the British Isles. MAFF, London, 75 maps (unpaginated).

MILES, E., SHERMAN, J., FLUHARTY, D., GIBBS, S., TANAKA, S. and ODA, M. (1982). Atlas of Marine Use in the North Pacific Region. University of California, Berkeley, Los Angeles and London, 103 pp.

An Assessment of the Relation of the Presentations to Exclusive Economic Zones

The EEZ: Opportunities Under the Legal Regime: A Summing Up

Patricia Birnie, London School of Economics

PRELIMINARY REMARKS

In 1974 the David Davies Institute published a study of the uses of the North Sea entitled "North Sea: Challenge and Opportunity". Seven years later the Greenwich Forum held a conference entitled "The North Sea: A New International Regime" at which I gave a paper. On January 1, 1977 a major change had occurred in UK law and policy: a new 200 mile fishing regime was asserted from that date but the David Davies Institute's call for a new approach to the challenges presented by the uses of the area's resources have not led to any new overall regime; and thus I entitled my paper "The North Sea: A Challenge of Disorganised Opportunities?." Britain appeared to be awaiting the end of the United Nations Law of the Sea Conference before adopting any further changes, including an Exclusive Economic Zone.

THE LEGAL REGIME

By 20 May 1986, as Professor Brown explained in his paper, the new Law of the Sea Convention (LOSC) concluded by the UNCLOS III on December 12, 1982, has provided a major opportunity for the creation of a new regime based on acceptance of a 12 mile limit for the territorial sea and acceptance and development of the concept of the 200 mile Exclusive Economic Zone (EEZ) and expansively revising the definition of the continental shelf. The EEZ consists of a new bundle of rights and jurisdictions. It allows coastal states to exercise sovereign rights over the living and non-living resources of the seabed, its subsoil and the water column above for the purpose of exploring and exploiting, conserving and managing the natural resources. It also recognises the coastal state's exclusive right to other economic uses of this zone such as the uses of waves, tides and currents for making energy and the creation of artificial islands (concerning the variety of prospects for which the SUT has already held a conference).

Some jurisdiction, not exclusive but subject to restraints laid down in other parts of the LOSC can also be exercised over scientific research and for purposes of protecting the marine environment. So far as the rights affect the seabed they are to be exercised subject to the provisions laid down in that part of the LOSC that concerns the continental shelf. The latter is now defined in the LOSC to include this 200 mile area of seabed, irrespective of the existence of any naturally prolonging extension of the land territory, although such extension beyond 200 miles can also be claimed, if it exists, i.e., if the geological or geophysical evidence supports such a claim. The LOSC makes it clear that such rights can now be claimed to the edge of the continental margin, though not beyond 350 miles from the base of its slope or 100 miles from the 2,500 metre isobath.

The LOSC is not yet in force and, therefore, not yet binding. However Professor Brown explained the historical background to these LOSC provisions and why parts of it, including its provisions on sovereign rights over fisheries and the continental shelf within the limits prescribed, can almost certainly be said to have become part of customary law. The UK has neither signed nor acceded to this Convention, however. Its current policy is based so far as seabed rights are concerned on the continental shelf doctrine that emerged after World War II and was codified in the 1958 Geneva Convention on the Continental Shelf to which the UK is still a party. The rights it permits have been asserted by the UK in the 1964 Continental Shelf Act and related acts and orders; beyond that the UK regards the 1958

High Seas Convention, codifying the doctrine of freedom of access, as applicable in the absence of other relevant conventions in force and accepted by it. So far as fisheries are concerned UK policy is based on the gradual extension that has occurred of coastal state's jurisdiction over fisheries culminating in 1976/77 in the assertion by the UK of jurisdiction, over fisheries only, within a 200 mile limit off its coasts in the 1976 Fishery Limits Act.

The UK, as a non-party to the LOSC, bases its current claims to jurisdiction over sea and seabed not on the LOSC but on the treaties and customs described above and such aspects of customary law as it regards as having developed. It is thus important at the outset to realise that the UK does not claim an EEZ as provided in the LOSC and, as Professor Brown pointed out, unlike the territorial sea or continental shelf such zones have to be claimed; they are not thrust upon states either by custom or treaty. Thus the UK at present has sovereignty (i.e. full powers and rights) in its internal waters and, subject to innocent passage of foreign vessels, in its territorial sea, which it still limits to 3 nautical miles, not the 12 miles approved in the LOSC. It claims an Exclusive Fisheries Zone (EFZ) of 200 miles but not rights over the seabed within that limit except to the extent that a legal continental shelf exists in conformity with the terms of the 1958 Convention as interpreted in customary law i.e. where there is a natural prolongation of the land territory into and under the sea. It does not claim the other rights within the 200 mile zone that the LOSC allows i.e. sovereign rights over other economic uses such as use of waves for energy etc., or jurisdiction over artificial islands, preservation of the marine environment or marine scientific research though it can exercise all these rights within its internal waters and territorial sea. It retains, as already stated, a 3 mile limit for the latter though it is contemplating extending it to 12 n.m. in the near future. It could claim all the rights in an EEZ permitted in the LOSC, even if it never becomes party to it or the Convention never enters into force, if on evaluation of state practice it decides that the full EEZ, not just the fisheries provisions, have been so widely accepted as to have become part of customary international law. The International Court of Justice (ICJ) recently decided in the Malta-Libya case that a 200 mile continental shelf, determined by distance alone, not geological or other factors, was already accepted as part of international customary law.

As far as resource opportunities are concerned at the present time current UK policies in relation to the seas and seabed are not restrictive, so far as the British Isles are concerned, since we have full opportunities to explore and exploit all fish and continental shelf resources and, as this conference has revealed are either not technically equipped or not otherwise commercially prepared to exploit wave energy etc. or to build artifical islands beyond our territorial waters. Within the scope of these qualifications we must, however, ascertain what precisely are the resource opportunities in UK jurisdictional areas and consider also whether in other areas of the world there arise other opportunities and benefits deriving from the assertion of EEZs as defined in the LOSC. I shall look at the suggestions put to us roughly in order that the speakers made them with the notable exception of Dr. Hill's bugle call to rally round the banner of the EEZ and march into a new era of opportunity for public and private enterprise in and under the seas that it symbolises.

SCIENTIFIC RESEARCH

Lord Sherfield told us that the House of Lords Select Committee on Science and Technology had considered that declaration of an EEZ would give us new opportunities, of which we ought to take advantage, to conduct scientific research throughout the world and participate in international programmes and organisations which would advertise the abilities of British scientists and the contribution we could make.

Unfortunately, however, the EZ is being used by those that have claimed one to control or prevent not to encourage foreign research within 200 miles of their coasts. That is what the LOSC permits. It requires that permission be sought of the coastal state to conduct research in its zone and prescribes that, in certain circumstances, for example where research is directed to resources, the coastal state can refuse such consent, though it must not refuse it "in normal circumstances", which unfortunately are not defined. Moreover, the LOSC allows coastal states to impose numerous requirements on researchers such as that a scientist from the coastal state be carried on board, that data be shared and interpreted and findings published. It will be much more difficult to put research expeditions into foreign waters in future in areas that formerly were regarded as high seas, open to freedom of research by our and other countries' scientists. I cannot see an assertion of an EEZ by us increasing our bargaining power for entry to foreign zones; we already participate in many international programmes and seem to have neither desire nor need to control foreign research beyond our own territorial seas. Our rights over the continental shelf based on the 1958 Convention already allow us considerable control over research directed to it. Obviously we should continue to make the most of such opportunities as <u>are</u> available in other states zones; the need of developing states to investigate the resources in and water quality etc. of their zones has obviously created new markets for our scientists, but we must not overlook the difficulties that also await us in reassuring such states that we are not endeavouring to take advantage of their lack of scientific expertise and equipment.

<u>RESOURCE DEVELOPMENT</u>

1. <u>MINERAL RESOURCES</u>

Dr. Cronan told us that the current position concerning various ocean minerals was as follows. Placers were technically and economically exploitable now and were in fact being exploited. It is up to industry to make the most of this opportunity. Phosphorite would be exploitable in the next century, which is not all that far away; exploratory plans can now be laid. Polymetallic nodules and oceanic crusts may be exploitable at the turn of the century. A couple of British companies have participated in multinational consortia which are developing the technical means and financial backing necessary to enable this industry to start operations. Meanwhile, as explained by Professor Brown, uncertainty concerning the legal regimes applicable inhibits investment and development. The UK supports the US and FRG (Federal Republic of Germany) in not ratifying the LOSC which establishes an International Authority to supervise and regulate such activities in the international area beyond the continental margin, though this approach has been endorsed by the 159 signatures attracted by the LOSC. The UK argument that the freedom of the high seas still extends to the seabed of the international areas and sanctions open access for all in the absence of acceptance by the UK and entry into force, of the LOSC has shaken confidence in the legal and, therefore, commercial viability of operating in this area. Nodules and crusts may thus be exploited only on and within the continental margins that are clearly accepted by all as being subject to the coastal state's jurisdiction. Metalliferous muds deriving from oceanic sinks in the sea floor are now subject to exploration of their economic potential but it is not yet known whether it will be technically or economically feasible to exploit them in the light of current data and knowledge. Seekers after opportunity will need to watch developments here in EEZs in relevant parts of the world.

2. FISHERIES RESOURCES

Both Dr. Gulland and Dr. Jones dealt with the range of exploitable fishery resources. After covering the nature and distribution of marketable species, the former indicated that there was still a lot of work to be done by fish economists, fish biologists and fish managers in order to get the best return from fisheries, especially from multi-species fisheries. There was scope also for marketing species new to the British markets which hitherto had been untried or met with consumer resistance. Marketing experts will need to set about changing or extending British fish-eating patterns and preferences. Some opportunities exist for exploiting distance water fisheries e.g. round the Falkland Islands and in the Southern Oceans generally, but they are few; most marketable species are found over the continental shelves or economic zones.

Dr. Jones explained the theories for ensuring a maximum sustainable yield of fisheries, which would help to maintain the productivity of stocks rather than their size. Productivity was affected by grazing, especially by availability of nutrients and food chain relations. Management formulae had to take account of these factors in calculating both how to sustain and to increase yields in the long term by development of a fishing strategy which would prevent stock collapses. Assessing stocks size was not a precise science but there was a variety of methods and sources and much scientific advice available.

There are now opportunities for the EC (European Community) to fill the past gaps in management of fisheries; this could lead to a higher TAC (Total Allowable Catch) being available in the long term within the EC's Common Fisheries Pond. I must point out, however, that the UK's existing EFZ is subject to the CFP and it would make no difference to such opportunities whether or not the UK converted the FZ into an EFZ; the latter also would be subject the CFP so far as fishing opportunities within it were concerned. But, of course, most of the global fisheries described will be inside the EEZ's or EFZ's of foreign states other than EC members and to gain access to opportunities in such zones agreements will have to be negotiated. There still remain, as indicated by Dr. Gulland, opportunities in some areas of the high seas but mostly in areas far from the UK, as in the South Atlantic. Here opportunities exist but present a major challenge to bold investment and new approaches following the collapse of most of the UK deepwater fleets after the advent of the 200 mile fisheries zone in Iceland, Norway and elsewhere.

3. RENEWABLE ENERGY RESOURCES

M. Marchand, in his review of the various methods of producing energy from various alteration marine based sources, painted a somewhat gloomy picture of the immediate prospects for entrepreneurs in this field since, despite some imaginative schemes, the technical, economic and practical difficulties presented by all schemes at this stage, except the Severn Barrage proposal, rendered them poor bets this century except for any daring investor willing to risk the very long odds against success. The Severn Barrage, being located in a river estuary, lies wholly within our internal waters and would not require adoption of either an extended territorial sea limit or an EEZ.

The opportunities in this area in the foreseeable future seem to be for inventors, for artists and draughtsmen with the skill to depict their schemes, for engineers to develop the technologies required at an acceptable economic cost, and for economists to assess their financial feasibility. Banks and market research analysts might also be attracted at some point. Finally, there would be pleasant opportunities for improving one's French by visiting M. Marchand to keep up with any new developments or further imaginative ideas in this field.

RESOURCE ASSESSMENT

1. Survey techniques

In this field there certainly seem to be vast opportunities. If Mrs. Worthington had any remaining daughters it would seem to be in her interest to put them into hydrography or geophysics, though it is appreciated that there may be certain practical difficulties in accommodating them on board the specialist vessels involved. The assistance required by most states, since few have the necessary vessels, skills and equipment for the task, in delimiting the baselines from which their new zones are measured, as Rear Admiral Morris graphically illustrated, has created a vast new worldwide market which Britain is particularly qualified to exploit. The R.N. Hydrographic Service is already contracting out its services to take advantage of this, but unlimited opportunities exist for private companies, suitably qualified and equipped. The conclusion of the LOSC, its approval of the continued and extended use of the straight baseline system, especially by the newly created "archipelagic states", and the attention these events have focussed on interpretation and application of the relevant provisions of the LOSC, including the formula for delimiting the outer limit of the "natural" continental shelf, has created many new customers for the services of hydrographers, geophysicists, cartographers and the builders of hydrographic survey vessels. It has to be recalled, however, that all the new extended limits are drawn from the baselines of the territorial sea; it is not necessary to claim any of the new limits in order correctly to locate the points from which these baselines can be drawn.

Dr. Francis confirmed the opportunities that also exist for development of precise navigational systems to enable vessels to operate within the new zonal or shelf limits. He reminded us that 59 EEZs, as well as many EFZs, have now been unilaterally declared, even before the entry into force of the LOSC. Thus these opportunities exist <u>now.</u>

There is, I should however point out, one opportunity the UK will lose if it has not become a party to the LOSC by the time it enters into force and that is to be appointed to the Continental Shelf Boundary Commission which will review and advise on the evidence concerning the outer limits of their shelves submitted to it by states party in conformity with the Convention. With the expertise so evidently available in the UK in this field, we would surely be in an excellent position to provide a Commissioner. The UK is not, however, precluded from providing specialist advisors both toparties and to the courts themselves - be they the International Court of Justice or arbitral tribunals - when hearing cases concerning delimitation of continental shelves between neighbouring states, and has done so. One retired R.N. Hydrographer advised the arbitral tribunal, for example, in the recent delimitation between Guinea and Guinea Bissau. There will be many more opportunities of these kinds as both Admiral Morris and Dr. Francis have illustrated.

2. Remote Sensing from Space

Dr. Thomas outlined the enormous amount of detailed information concerning the oceans and ice-covered areas that it was now possible to garner using satellites; not only on the surface visibles including ocean colours and seas surface resources, but invisibles such as the sea surface temperature. Satellites can carry a wide variety of technical equipment for such purposes. Such equipment is highly versatile and can be used for a variety of purposes - identifying fishing opportunities; weather and wave prediction; pollution detection; ships' and icebergs' movements; tuna and salmon migrations; global productivity to name but a few. This would seem to provide so many valuable services to so many industries and governments worldwide that opportunities to manufacture, market and otherwise develop the equipment to put into satellite, if not the satellites themselves, abound. Undoubtedly the adoption of EFZs, and even more so of EEZs, will stimulate demand for such information as states seek to maximise the benefits of both the resources and the enforcement jurisdictions newly available to them in such extensive areas. Should the UK adopt an EEZ no doubt home-based industries and appropriate government departments will display similar interest but clearly opportunities already exist for the defence and commercial interests involved in scrutinising them in relation to the ECs common fish pond, to foreign EEZs and the high seas areas beyond.

We must, however, remember that there may be disadvantages in using, if not in obtaining, too much information on fish movements at a time when there are needs in relation to particular stocks, especially of salmon and tuna, to restrict catches in the short-term, in the interest of conservation, following depletion of such stocks consequent upon intensive overfishing.

3. The UK and its Dependencies: Prospective EEZ Opportunities

Professor Smith's survey of the extent and richness of the seabed areas round the UK and its overseas dependencies indicated that there was now sufficient information available following UK public investment in research into the breadth and resources of the shelves in these areas, to make it worthwhole to extend and actually exercise UK jurisdiction over larger areas. In the light of this research, in his view, the UK should declare an EEZ not only round the British Isles and Rockall, where there is sufficient knowledge (which could quickly be added to) to justify such as extension, but also round the seventeen British overseas dependencies ranging from Bermuda, Tristan da Cunha and the Falklands to Pitcairn Island.

We do, of course, already have continental shelf rights in such areas to the extent that a shelf exists in terms of a natural prolongation of the land territory. We could legitimately also claim fishing zones, if we have not already done so (although in some cases such as the Falkland Islands and Gibraltar there are political difficulties inhibiting this), but neither of these options would give us here, as they do round the British Isles, advantages comparable to the EEZs so vigorously advocated by Professor Smith, since there is only a small area of natural underwater prolongation of the land territory off most dependencies. An EEZ would enable us to claim sovereign rights to exploit 200 miles of seabed even in the absence of any "natural" shelf. This would indeed open up new areas for such opportunities as exist therein, though oil and gas resources are more likely to be located in natural continental shelves.

OPPORTUNITIES ARISING FOR AND FROM MANAGEMENT OF THE EEZ

1. Managing the Resources: Japan, A Case Study

Daniel Spagni outlined the positive arrangements for the promotion of research and development , especially of technology relating to exploitation of resources in Japan. Japan's view was that this policy was a good investment that would pay dividends. It had therefore invested very large sums indeed in a programme of positive development of the technology to exploit its own EEZ fully and effectively. It aimed to increase, inter alia, its oceans and fish catches; aquaculture and marine ranching.

The Japanese programme model would seem to provide an attractive and encouraging model for UK development although we must remember that Japan is not nearly so advantageously situated in terms of natural shelf and EFZ are as is the UK and thus such levels of investment may relatively be more worthwhile in Japan. Nonetheless the positive and dynamic aspects of such efforts do require that we re-examine our own policies towards our existing and potential resource areas.

2. The Marine Atlas

The Marine Atlas, the contents of and difficulties in compiling which Mr. Ramster's outlined, provides a variety informative maps which offer a useful guide to opportunities for those willing to investigate the resource opportunities recorded and illustrated in them. These range from oil and gas and fishing to sand and gravel and dumping sites. This paper was a useful complement to Professor Smith's and others so far as UK marine resources are concerned. We cannot in the light of the Atlas' existence say we do not know what opportunities lie on our own doorstep.

3. The United States' EEZ

I have saved the day's most invigorating paper until the end of this resume, though chronologically it was delivered after

Dr. Jones at the beginning of the session on EEZ opportunities, because I am sure that all would agree that it was electrifying in its enthusiasm and drive. I refer, of course, to Dr. Hill's remarkable paper on the extent to which and the speed with which the United States' consciousness of ocean opportunities has been raised by the declaration in 1983 by President Reagan of an EEZ. It appears to have revived the US's pioneer spirit by setting the challenge of a new frontier of vast extent. National interest in investigating it has been stimulated as has general public education covering its resources and the opportunities for multiple uses. Crises, such as the OPEC oil price rise, diverted the US's attention, as it did ours, to home-based fuel resources and prompted the release of funds to develop its offshore oil and gas resources in particular, the building of new research vessels, and development of new technology, such as GLORIA, to investigate the ocean floor. A better, less bureaucratic infrastructure has, it seems, also emerged since the EZ declaration to administer the new comprehensive legislation directed at minerals other than oil, as well as the oil and gas themselves.

Despite the unforseen problem, identified by Dr. Hill that much of the resultant information remains classified because of its military implications, enough is available to reveal the myopia of the old approach directed exclusively to the oil industry. The development of a more streamlined resource management approach has been of immense benefit to the United States which aims at self-sufficiency in oil and minerals in an attempt to safeguard itself against adverse changes in the world political system which might lead to the withholding of strategic minerals from the US. It expects national development now to focus on the tools to develop mineral crusts and nodules within its EEZ area to help it to achieve this aim, though this is a very long term aim since 40 to 60 years and vast infusions of cash will be required to meet it.

This address was inspiring both in its enthusiasm and the verve with which it was delivered and Dr. Hills implication clearly seemed to be that we could achieve a similar rise in the levels of the British public's interest in its offshore opportunities if we too declared an EEZ. But we must pause and ask a few questions concerning whether, if this dynamic policy is to be transfused into the bloodstream of British Government in order to reinvigorate its policies, it is necessary in fact to declare an EEZ. It seems to have had great symbolic significance in the US; the US government and people have been inspired with a sense of acquisition of new territory which must be developed by a national programme just as our governments and people were in the days of Colonial Empire.

But we must remind ourselves that though the US has a vast area of outer limits open to the deep oceans and bounded by no other state the UK borders several virtually semi-enclosed seas - the North, Irish and Celtic Seas and the English Channel - areas which are amongst the most densely used and transited in the world. Moreover, the US EEZ is not a true EEZ as provided in the LOSC. Only some of the rights prescribed in the LOSC are included in it and part of the US objective in declaring the zone was to establish a regime that would deviate from the LOS treaty, which it was rejecting in order to undermine the parts relating to the deep seabed regime. The US also has problems with federal and State relations which the new EEZ regime might help to resolve, a problem which is not shared by the UK. Above all the UK already has the resource rights claimed by the US in its EZ in the form of rights over the extensive UK shelf and in the UK fishing zone. The US, moreover, does not have a straight baseline system whereas we do have such a system off the West Coast of Scotland round the Hebrides, which gives us considerable advantages in extending our control over the shelf and fishing zone west and north of these islands.

Whether or not the day's papers have established that we

need an EEZ to realise the full potential of the areas off our coasts, these proceedings and the excellent and detailed range of papers on all aspects of that potential, have surely served to make it clear beyond doubt that a large number of opportunities for industry, individuals and government departments do exist in these areas as well as in the EEZs of other states and that such opportunities need to be seized more vigorously than they have been to date. Dr. Hill told us that jobs had been created for many FTEs - Full Time Equivalents - as a result of the new US EEZ policy, FTE meaning in US jargon a person. It will be a pity if, after the windows opened by today's programme on an era of new opportunities, the policy so often adopted by many of those involved in resource development in Britain continues to be the one of FTE - in British jargon Full Time Equivocacy.

EXCLUSIVE ECONOMIC ZONES

Report from the Rapporteur

Introduction

As Dr. Birnie's paper assesses in detail the significance of the conference presentations to Exclusive Economic Zones, this report focuses upon major themes and issues which recurred throughout the conference contributions. This is approached by highlighting certain aspects which emerged in the opening session, followed by consideration of the three major conference themes of development, assessment and management.

The Setting

The opening session was mainly concerned with three things: new ideas (Chairman), the first practical moves in the development and management of marine resources (Lord Sherfield), and the legal basis upon which rests the next phase in the history of the EEZ (Professor Brown). The EEZ is itself a relatively new concept which in turn is a reflection of global, philosophical and cultural trends, not least in the fields of science and technology, and which has provided new opportunities for both exploitation and management. In practical terms this is reflected in the recent House of Lords report on marine science and technology, which emphasised the need for action in environmental protection, resource evaluation and underwater engineering. The developments in the international law of the sea since the 1950s underline these circumstances, emphasising the central role and responsibilities of the coastal state as evidenced, for example, in the EEZ provisions. This will involve complex situations in many areas of resources development and management, not least in the allocation of sea space. The division of the seas around Rockall are an example of such allocation problems in a UK context.

Development

The development theme at the conference was primarily concerned with outlining the nature of extractive marine resources, the three main papers dealing respectively with mineral, fisheries and renewable energy resources. These in turn highlighted other themes which have more to do with the development process itself, bearing in mind that much in this field lies in the future rather than the present. Development is essentially concerned with the human element which provides both push and direction, not only through the expansion and application of science and technology, but also providing the means of control through legal frameworks and management organisations, objectives and techniques.

A number of important points emerged. First, despite the lack of detailed surveys of most of the world ocean (including the EEZ), the knowledge of extractive resources is both substantial and rapidly increasing, as, for example, in the minerals field. Furthermore, this emphasises the resource potential of the EEZ as the major source of extractive resources in virtually all fields. Due to the emphasis on resource extraction, the important area of environmental resources was scarcely mentioned; the marine environment as a whole is used for transport and communication, strategic purposes, recreaction, research, conservation and waste disposal. Nor were the intensely used parts of the EEZ adjacent to the coasts covered in any detail.

The increasing knowledge of the EEZ and its resource potential in turn focused attention on the management potential, emphasised in the fisheries paper. The importance of continuity and progress already made emerged here, note being taken of the good work accomplished by the fisheries commissions before the advent of the EEZ. Not explicity covered, however, was the geographical dimension of development on the global scale, where use intensities and environmental impacts are greatest in the industrial core region of the global economy centred on the North Atlantic and the fast growing periphery of the North Pacific rim, where most of the ocean management infrastructure is also located. Beyond lies the developing tropical and southern peripheries, dependent upon the core for both development and provision of management resources. Some of the opportunities here were returned to in the final session, notably with regard to oceanic island EEZs.

Assessment

The assessment theme is concerned with the initial technical challenges of resource evaluation, and the conference papers dealt with mensuration, monitoring and resource assessment *per se.* The two principal areas covered in the mensuration theme were the survey work underlying delimitation of base lines upon which spatial allocation among states is based; and geophysical surveying which is the first step in mineral resource evaluation. The monitoring theme emerged most clearly in the remote sensing and fisheries contributions where research effort starts with building reliable data bases for both extractive resources and the environment as a whole. The considerable potential at the remote sensing technological frontier is notable. The complexities of resource assessment beyond the initial survey and monitoring activities were well illustrated in the fisheries contributions, but also apply to other extractive resources in various ways.

Two further themese which emerged concerned information technology and time scales. The need for continuing development of information technology was highlighted in the remote sensing contribution, in the generation of immense quantities of data in this field, which require computer storage and analysis, and will probably require the development of the next generation of intelligent machines to be properly used. The importance of long-term horizons in resource assessment was emphasised in the fisheries contribution. This is a central concern in assessment of renewable resources and environmental impact, but is also significant in the non-renewable resource field.

Management

The management theme revolves around decisions based upon objectives, not ONLY THE OBVIOUS ones stemming from govermental responsibilities, but also those made by "developers" - the sea users. Here the papers focused upon specific case studies - principally the UK, USA and Japan, and upon the role of information, exemplified by the marine resource atlas, and ending with an assessment of the relation of the conference contributions to EEZs. The first major theme emphasised was the need for public awarenexx and participation in maritime affairs, which emerged clearly in the US situation, where maritime research and administration budgets are under pressure. Following from this are the circumstances surrounding provision of information. There remains a lack of information related to the need for more research; a tendency towards classification of much of what is available - the US case was again cited here in a strategic context, although it applies also in a commercial context; and the importance of the users of this information. The marine resource atlas is seen as a necessary first step in the provision of information for both public and specialised users, and the considerable progress made in the industrial North Atlantic and North Pacific was reviewed.

A third area of concern is that of national policies and plans for the marine environment, which derives from the key role of the coastal state referred to in a fisheries context in the opening session. It appears that Japan is furthest ahead in the formulation of the perceived need for integrated national policies in this field. The very least that is required is the proper co-ordination of marine affairs. This in turn highlighted yet again the theme of time scales, and the need to adopt the long view in maritime affairs as far as the exploitation of the EEZ is concerned. The final contribution returned in part to the UK focus and the need for the development of the UK "EEZ" referred to in the keynote address. A disorganised approach remains characteristic.

Conclusion

In reviewing the contribution of the conference to the EEZ, the first contribution is the organised approach to thinking about it: the development theme which can contribute so much to understanding the pressures for exploitation; the resource assessment theme concerned with the technology for evaluation of the EEZ; and the management theme concerned with objectives and decisions contributing towards the control of the first two sets of processes.

A second major contribution is the identification of common practical concerns which cut across the fields as covered by the individual papers. This concerns the continuing work in the establishment of a legal framework in the wake of the 1982 Law of the Sea Convention; the need for continued development of research and the information technology to cope with the results and their application, the basic fields including survey, monitoring and assessment programmes; and the development of national policies for EEZs, including management frameworks and proper, accessible data bases for sea users and policy makers.

A third and final theme which emerged throughout development, assessment and management is the need to take the long view with respect to the stewardship of the EEZ, not overly influenced by short-term economic and political considerations. This is especially necessary in what remains a new and challenging frontier - in international law and relations, industrialisation and environmental management. In this it is necessary to maintain a global view, encompassing both industrial and industrialising countries, reflecting the oneness of the world ocean itself.

H D SMITH

EEZ IN PERSPECTIVE

Preliminary Impressions of the Conference

E.F. Roots, Department of the Environment, Canada

Ladies and gentlemen, it is my final duty as chairman to pass along to you some informal comments on the presentations that we have heard and the discussions we have participated in today, and to ask you, in turn to make your own comments as we digest our ideas and relax after a stimulating day and a good dinner. In so doing, I shall not attempt to match the perceptive and thought-provoking assessment by Dr. Birnie, who gave us most valuable insights, particularly from an economic and legal perspective, into the linkages between and the significance of the many aspects of Exclusive Economic Zones that we have touched upon. Nor can I add anything to the masterful summary by our rapporteur, Dr. Hance Smith, who was able to tell us, in clear and simple terms, what it was we had been talking about all day. My job as it has been given to me just now by our indefatigable organizers, is to share with you some initial reactions about what the conference has meant in a broad scientific and policy context. I can do this only without advance preparation and from the perspective of one who has been sitting in the chair and listening - with, albeit, one eye on the clock. At very short notice, I have attempted to sort out my impressions under a number of headings.

1. Timing

One of the first clear impressions is that the timing has been right to have a thorough, multi-subject and international discussion on this subject. Lord Sherfield pointed out this morning that the government of the United Kingdom is receptive to a clear message, from a broad range of knowledgeable interests, on the consequences of making, or not making, a number of important decisions about exclusive economic zones; and that the appropriate time to deliver that message is in the next few months. The situation is similar in several other countries. By count of numbers, the world is about halfway along to ratification of UNCLOS-III, and this is an appropriate time, now that there is a little experience and clarification of both the problems and potential opportunities, for governments to consider from many points of view what do the EEZ concepts really mean in their various scientific and environmental as well as their political and economic ramifications. It is also appropriate for governments to consider what technical data and scientific information should they have in order to make the best political decisions. A large number of regional problems that are coming to a head in the very near future call for better understanding and appreciation of the issues we have been discussing today:- I draw to your attention questions in the Arctic Ocean, the Caribbean, the western Pacific, and sub-Antarctic waters as examples, and each of you can add your own examples. All of these require a marrying of science, surveying, and economics to ocean policies in very short order.

But there are not only problems. There are opportunities in the next few months to use better knowledge of marine areas within a few hundred miles of land to improve the whole human situation. These opportunities also mean that the time is very appropriate to examine all sides of EEZ issues. For example, the World Commission on Environment and Development, established by the United Nations General Assembly to look deeply into the fundamental causes of societal and international problems as they are rooted in environmental, resources, population and policy issues, is about halfway through its information-gathering phase. Among the most important of the recent changes in the way that humans are managing the planet are the changes related to EEZ; surely it is most important that the clearest and most balanced information about EEZ issues, in a global and not a narrow national context, be brought before the World Commission as soon as possible.

There are other timely opportunities: for example, the International Council of Scientific Unions' Committee on Data for Science and Technology is for the first time giving concerted international attention to the standards and the management of data on the environment; in fact an international workshop on the subject is being held this very week [1]. From the issues we have been discussing today, it is clear that reliable information and data, accessible through integrated or compatible data management systems, on oceanic characteristics is going to be of increasing policy and economic importance. The political trends would appear to be fragmenting the responsibility for and "ownership" of such data; in many cases the responsibility is passing to national authorities that are not equipped with sophisticated technology to obtain or manage it. At the same time, remote

1. ICSU/CODATA Workshop on "Directions for Internationally Compatible Environmental Data", Montreal, Canada, 19-23 May 1986.

sensing technology is revolutionizing the capability, for those who can afford it, of surveying the whole ocean; and international ocean data and global environmental monitoring systems are being established. It is clear that the scientific information and data aspects alone of EEZ's make a discussion and clarification of the issues very timely. Among other uses, comprehensive data on world ocean characteristics and resources, especially in the EEZ areas that include the overwhelming preponderance of exploitable fish stocks and seabed hydrocarbon deposits, is essential to a proper study of "global change" which is the burden of the ICSU International Geosphere-Biosphere Programme [2], whose planning is well advanced and which is likely to be a dominant theme of international natural sciences research over the next couple of decades. So considerations of what EEZ's mean for science right at this time are very apposite.

2. Balanced Uses of the Oceans - or International Turmoil?

As I sat listening today, I found myself pondering on whether the developments that were taking place with regard to EEZ's, and the knowledge we were obtaining about marginal parts of the oceans, were helping to develop national positions that take into account the five functions of the oceans that relate to human use that I referred to this morning. It seemed to me that the record so far is quite mixed. The declaration and implementation of EEZ's raises new problems for many nations as well as helping meet some political goals and, perhaps, promising economic benefits. With all these changes and issues, we may be embarking on another period of turbulent national and international ocean management similar to that of the seventeenth century.

3. Responsibilities of Coastal Nations

A third and sobering impression left by the day was that the new situations signalled a significant change in the responsibilities of coastal and island nations relative to inland nations. Despite attempts in the United Nations and by treaties or philosophical treatises to recognize legal equality between nations, different states have never been equal in opportunity or responsibility, as any country that is lacking in petroleum resources, or that lies along a natural trade route, or which is subjected

---

2. Global Change in the Geosphere-Biosphere: Initial Priorities for an IGBP. U.S. National Research Council, 1986; and numerous ICSU publications.

to severe climate change knows only too well. But the extension of economic jurisdiction of coastal states over large parts of the ocean increases the international and global responsibilities of countries that by geographical accident happen to have convex coastlines. This is no idle or academic point. In the old days of mare clausem, when for example in the eleventh century the bishopric of Hamburg included the North Atlantic and the lands within it and on either side of it, or some centuries later when under mare liberum Britannia ruled the waves but others had access to them, the responsibility could be discharged, mostly unilaterally, through a broad policy and a reasonable budget for the navy. But today, the 103 coastal states that have declared EEZ's or their equivalent have taken unto themselves a much more complex task which includes, under a number of other United Nations declarations, responsibilities to inland states for managing the common heritage of mankind and protection of a large part of the world's environment. These responsibilities change the internal responsibilities and will undoubtedly affect the priorities and internal structure of agencies within coastal countries; they shift the attentions and methods of operations of industry; and they clearly change the scope and emphasis required of a country's scientists and technologists. We have heard Dr. Hill describe how even in a country as sophisticated and with as many varied interests as the United States, there has been a strong psychological response by the public at large to the thrill and responsibility of "owning" a new piece of ocean. This change has not of course gone un-noticed by the land-locked states, and we can probably expect to hear more about it at the United Nations in the next few years.

4. Law and Geography

Another point that lay beneath several issues discussed today was the degree to which EEZ and UNCLOS issues were bringing about legal definitions for natural features and processes. Throughout much of the development of the natural sciences, it has been a tenet of the attempted "objectivity" of science that natural features were described and classified on bases that were as independent of human uses or institutional definitions as possible. There have been exceptions of course: "mean high water line" became a legal term on charts that only approximated where the water got to; a whale is legally a fish in some countries, its mammalian biology notwithstanding; and a 50-year flood is an insurance calculation to which the river and climate may pay little attention. But through the protracted discussions of the Law of the Sea, and the Exclusive Economic Zone debates, the lawyers have largely re-defined physical bathymetry. I have spent many years on field surveys of continental shelves; but as a geologist with some knowledge of oceans and geophysics I defy any other geologist to read Article 76 of UNCLOS III without a frown of puzzlement or a smile of incredulity. The strange thing is, the definitions work, after a fashion; and I suppose we earth scientists will get accustomed to man-made legal definitions of important parts of our planet which took millions of years to evolve. Another demonstration, perhaps, of the essential interdependence of man and nature. Or is it that with the declaration of Exclusive Economic Zones the lawyers and God are now becoming indistinguishable?

5. The Need and Opportunity to Synthesize Science and Policy

No one who was present at this conference could fail to be impressed by the circumstance that in nearly every discussion, geography, technology, oceanographic sciences and policy questions were all mixed. The resolution of issues surrounding exclusive economic zones were seen to depend largely on how well a nation was able, if I may use the vernacular, to "get its act together". We were presented with some interesting comparisons between Japan, the United States, and the United Kingdom in this regard. Listening as a Canadian, from a country that has made some unilateral moves in this field but which is not immune from difficulties in matching its science to its policies or its policies to its science, and which has in its national coat of arms the motto "A mari usque ad mare", but which says nothing about how far out to sea we go, I found myself thinking how difficult it is even for a country with some experience in this game to co-ordinate its policy development with oceanographic information and technology development. If I had been a delegate from Malta or Tonga or Indonesia, what would be my problems? However, one of the positive aspects of the whole EEZ discussion is that it provides a spur to synthesize our science and technology, and to examine its relationship to policies and economic planning. If this is done first for offshore and coastal areas we will at least have some national science policies for those regions that relate both to broader national science planning and to international science and planning. Thus, although the EEZ movement is essentially a nationalistic one, one of its developments and consequences might be to achieve greater internationalization of science and science policy.

This broader view of the scientific consequences of implementation of

EEZ's may become particularly apparent with respect to resources management and environmental protection. Questions of the location and identification of seabed minerals, and the technologies to extract them, can hardly be addressed separately in each national Zone, and we are already witnessing the paradox of countries having to collaborate scientifically in order to obtain the knowledge to develop separate national policies for resources under their exclusive jurisdiction. The same is equally true with regard to fisheries and will become even more so when, as we all hope, there will be progress toward rational international management of migratory fish stocks. As for maintenance of the quality of the marine environment, Part XII of the United Nations Convention on the Law of the Sea is quite specific on the need to increase scientific knowledge relevant to the marine environment and to use that knowledge in national policies in recognition of international environmental responsibilities. Articles 194 and 197 to 200 set forth some principles for the application of science to marine protection policies that would be hard for any state to disagree with, regardless of whether or not it declared an EEZ or signed UNCLOS-III.

6. Information Handling and Presentation

Many of the questions surrounding EEZ issues relate to scientific and technical information, its accessibility, reliability, and completion. Behind all these, but often ignored or taken for granted in policy and science considerations, are the questions of the systems and means for handling and presentation of appropriate data. As the issues change, become broader or more focussed, not only does the need for information change but there are also changes in the technologies and possibilities for assembling, archiving, and presenting the information. Mr. Ramster showed us this afternoon some of the problems and successes in incorporating new and changing information in the *Atlas of the Sea*.

Any useful information system, whether essentially graphic, like an atlas, or electronic, like a modern data bank, or literary, as with libraries and catalogues of abstracts, must be systematic and consistent in the way it classifies or presents its information across its subject range and over a period of time. Therefore, any changes in information handling or presentation must take place in steps, preferably at not very frequent intervals. Not only must the basic information technologies be selected in advance and more or less "frozen" for a useful period, but the questions that the users are likely to ask and their priorities for information must to a degree be decided upon by the "information people" even before the users themselves have given firm indication of what they need, or in many cases, before the full range of users has been identified for a dynamically changing, issue-laden subject like Exclusive Economic Zones.

In some cases, the lead time for decisions on data gathering or systems of presentation are quite long, and the decisions on what to measure, record or portray must be made - and estimates of cost made and resources committed - long before the information appears in a systematic way. Decisions on what sensors to include on a satellite and to what detail the information should be gathered require commitments 5 to 8 years before the satellite is launched. As we have heard, it takes at least ten years to bring about a major revision in an important atlas, and changes in world data systems are made infrequently and with considerable reluctance. For a rapidly developing multi-discipline field like EEZ's, this is a difficult problem indeed.

Mr. Ramster's talk made it very clear that the decisions on what to include and what to leave out from an atlas, and how to present it, cannot be a passive or totally responsive one. The setting up of an atlas requires positive decisions on what is important - in his case, with regard to oceanographic, hydrographic and resources knowledge. And what is included in the atlas influences what others think must be important. The World Data Centres for Oceanography likewise have a continuing problem of maintaining consistency of classification, so that researchers can obtain consistent data sets, while at the same time introducing new or revised categories that properly present data from new researches in fields that did not exist when the categories were established. The increased attention being given to EEZ's around the world by many agencies who previously were not part of the ocean information system seems to exacerbate these problems.

7. World Science and EEZ

As our conference was focussed on issues of Exclusive Economic Zones directly, discussions of science were concentrated on what do we know about EEZ's themselves, or on the scientific base for development of technologies that may be employed within EEZ's. We mentioned on several occasions, as noted above, that the issues surrounding EEZ's and the attention they were getting should serve to stimulate oceanographic science and its relationship to policy or data management. But many of us were conscious, I think, of another dimension; that the progress of world science and the increased understanding of natural or planetary processes is very important to EEZ's and their management. This is another area where the timing of our conference has been most appropriate. We are at one of those stages in world science where both theoretical understanding and technologies for obtaining information have recently surged ahead, and we are entering a period of critical analysis, synthesis, interpretation of what our new knowledge means and introspection of its significance. I need only remind you of the changes in our scientific thinking brought about in the last decade by plate tectonics, the discovery of sea-floor "smokers", of the mechanisms of genetic coding and evolutionary surges and extinctions, of biological modulation of planetary climate and chemistry. Or of the new data that are now available to us on a routine basis, such as satellite sensing of ocean surface temperature or the three-dimensional thermal structure of storms, and computer techniques to make these data useable in ways that researchers of the past could not have dreamed of. These advances in world science, and the approaches and techniques of science, have meant that a different knowledge background is available for discussions of EEZ's than was available for any previous discussions of ocean issues. Dr. Thomas drew our attention this afternoon to the wealth of data that are now available in some areas. We all know that there are several steps before data become information and information becomes knowledge; but the fact remains that the science base regarding oceans, planetary processes and biology has expanded spectacularly during the same period that EEZ's have become an issue.

One aspect of this expanded science base is a broadening of the understanding of the linkages between oceanographic phenomena, which are important to EEZ's directly, and other regional or global events. A good illustration of this was the chain of events and processes related to the El Niño water mass anomaly off the coast of Peru in 1981/82. Previous occurences of this anomaly, although not as spectacular, have had important effects on the South American anchovy catch -- surely they were important to the EEZ's concerned, but although the cause was understood to be not local but related to oceanographic and climatic conditions in the western Pacific, the events were of largely scientific interest outside the immediate area affected. During the most recent El Niño, however, global information systems have given us a picture of the anomalous local warming in the western Pacific, the trans-oceanic travel of the heated water pulse, the breakdown of water structure off South America, and of the effect of displaced water masses and heated air masses on the ocean currents and weather patterns in the North Pacific and eventually across North America and Europe. The South American anchovy fishery, already weakened by over-fishing, collapsed; as the main use of anchovies was as a protein additive for European cattle food, the prices of soybeans, which are the most suitable substitute proteins, rose in Europe and North America; however growing conditions in these areas were disorganized by anomalous floods and droughts that themselves were an aftermath of the El Niño disturbance. The consequent effects on world agriculture and international economic relations because of a natural disturbance in a Peruvian EEZ provide a dramatic example, not only of the inter-relationship of Exclusive Economic Zones to social and economic developments in the rest of the world, but also of the need for a world-wide science and information system to manage EEZ's and understand their role in human events.

Many major research programs that are building upon, and adding to, this newly acquired scientific capacity and global view, are international, multi-discipline, and include the oceans in their study of planetary or regional phenomena. As such, they will naturally include EEZ's of many countries in their areas of investigation. I have already referred to the International Geosphere-Biosphere Programme of the International Council of Scientific Unions (ICSU). Another ICSU-sponsored program is the Ocean Drilling Programme (ODP) which was mentioned earlier today. The International Institute of Applied Systems Analysis (IIASA) is developing a multination study on "Sustainable Development of the Biosphere" which is examining how human actions can be made compatible with a productive biosphere. The success of the World Climate Programme of the World Meteorological Organization is dependent on adequate complementary information from the oceans and research on oceanic/terrestrial conditions in many parts of the world. All these and other comprehensive studies will require that as Exclusive Economic Zones are put into effect, not only must the need for open and co-ordinated scientific study in all zones be accommodated, but it is necessary that recognition of the importance of such researches, from both a global and a national point of view, be incorporated in EEZ policies.

8. EEZ's and Developing Countries

Another point that seems pertinent, and important to mention as we review the conference, is that despite the talk of sophisticated science and the concerns of the maritime powers, Exclusive Economic Zones by no means belong to the rich nations. There were 139 sovereign coastal states, according to Lord Sherfield's count, who have the opportunity or responsibility to declare an EEZ or to refrain from doing so; and of these the majority are what in United Nations parlance would be called "developing countries". Many are countries that need financial and technical aid from the more heavily industrialized countries before they can get into a position to become economically self-sustaining in an increasingly technology-dependent world. Most of these countries are small. Their "use" of the ocean area within the Economic Zones that they have proclaimed or may proclaim for their exclusive use is not well developed, nor do they have the science information systems and capacity to obtain data or develop the technologies and management systems appropriate to their own EEZ's or marginal seas.

The problems faced by many developing countries in managing their EEZ's for sustained economic yield and environmental protection are truly great. Some of these were highlighted at a recent United Nations seminar in which I was priviledged to take part [3]; where the difficulties inherent in establishing a reliable hydrographic service, in many coastal countries that do not yet have one, were clearly apparent. It seems likely that one predictable result of the current trend toward declarations of Exclusive Economic Zones around the world will be a growth in the demand for overseas aid or country assistance programmes directed toward assistance in development of ocean resources and management of the marine environment. With a few exceptions, aid programmes until now have been focussed on terrestrial problems and opportunities; and in most donor countries the expertise and experience within the aid agencies is terrestrial or coastal. The declarations of EEZ's, and the problems and opportunities they raise for developing

3. United Nations; Inter-Regional Seminar on the Role of Surveying, Mapping and Charting in Country Development Programming; Aylmer, Quebec, Canada, 4-8 November 1985; proceedings available from U.N. Department of Technical Co-operation for Development, or from Surveys and Mapping Branch, Department of Energy Mines and Resources, Ottawa Canada K1A 0E4.

countries, can also be expected to cause changes in the aid programmes and overseas assistance agencies of developed countries.

9. Conclusion

It is clear, ladies and gentlemen, that we have not disposed of any issues related to Exclusive Economic Zones at this conference. But we have brought together, and examined in a larger context, many aspects of one of the most important recent developments in human world history. Where these developments will take us, none can say at this time; but it may not be too presumptuous to say that frank, informed discussions such as we have had today may play some small part in making the subsequent history more favourable both to mankind and to Nature.

Batch number: 10584448

Printed by Printforce, the Netherlands